Towards 4G Technologies

WILEY SERIES IN COMMUNICATIONS NETWORKING & DISTRIBUTED SYSTEMS

Series Editor: David Hutchison, *Lancaster University, Lancaster, UK*
Series Advisers: Serge Fdida, *Université Pierre et Marie Curie, Paris, France*
                   Joe Sventek, *University of Glasgow, Glasgow, UK*

The 'Wiley Series in Communications Networking & Distributed Systems' is a series of expert-level, technically detailed books covering cutting-edge research, and brand new developments as well as tutorial-style treatments in networking, middleware and software technologies for communications and distributed systems. The books will provide timely and reliable information about the state-of-the-art to researchers, advanced students and development engineers in the Telecommunications and the Computing sectors.

Other titles in the series:

Wright: *Voice over Packet Networks* 0-471-49516-6 (February 2001)
Jepsen: *Java for Telecommunications* 0-471-49826-2 (July 2001)
Sutton: *Secure Communications* 0-471-49904-8 (December 2001)
Stajano: *Security for Ubiquitous Computing* 0-470-84493-0 (February 2002)
Martin-Flatin: *Web-Based Management of IP Networks and Systems*, 0-471-48702-3 (September 2002)
Berman, Fox, Hey: *Grid Computing. Making the Global Infrastructure a Reality,* 0-470-85319-0 (March 2003)
Turner, Magill, Marples: *Service Provision. Technologies for Next Generation Communications* 0-470-85066-3 (April 2004)
Welzl: *Network Congestion Control: Managing Internet Traffic* 0-470-02528-X (July 2005)
Raz, Juhola, Serrat-Fernandez, Galis : *Fast and Efficient Context-Aware Services* 0-470-01668-X (April 2006)
Heckmann: *The Competitive Internet Service Provider* 0-470-01293-5 (April 2006)
Dressler: *Self-Organization in Sensor and Actor Networks* 0-470-02820-3 (November 2007)

Towards 4G Technologies

SERVICES WITH INITIATIVE

Edited by

Hendrik Berndt
DoCoMo Communications Laboratories Europe GmbH, Germany

John Wiley & Sons, Ltd

Copyright © 2008   John Wiley & Sons Ltd, The Atrium, Southern Gate, Chichester,
West Sussex PO19 8SQ, England

Telephone (+44) 1243 779777

Email (for orders and customer service enquiries): cs-books@wiley.co.uk
Visit our Home Page on www.wiley.com

Other Wiley Editorial Offices

John Wiley & Sons Inc., 111 River Street, Hoboken, NJ 07030, USA

Jossey-Bass, 989 Market Street, San Francisco, CA 94103-1741, USA

Wiley-VCH Verlag GmbH, Boschstr. 12, D-69469 Weinheim, Germany

John Wiley & Sons Australia Ltd, 42 McDougall Street, Milton, Queensland 4064, Australia

John Wiley & Sons (Asia) Pte Ltd, 2 Clementi Loop #02-01, Jin Xing Distripark, Singapore 129809

John Wiley & Sons Canada Ltd, 6045 Freemont Blvd, Mississauga, ONT, L5R 4J3, Canada

Wiley also publishes its books in a variety of electronic formats. Some content that appears in print may not be available in electronic books.

Library of Congress Cataloging-in-Publication Data

Towards 4G technologies : services with initiative / Edited by Hendrik Berndt.
    p. cm.
    Includes index.
    ISBN 978-0-470-01031-0 (cloth)
  1. Wireless communication systems.   I. Berndt, Hendrik.
    TK5103.2.T68   2008
    621.384 – dc22

               2007039300

British Library Cataloguing in Publication Data

A catalogue record for this book is available from the British Library

ISBN 978-0-470-01031-0 (H/B)

Typeset in 10/12pt Times by SNP Best-set Typesetter Ltd., Hong Kong.
Printed and bound in Great Britain by Antony Rowe Ltd, Chippenham, England.

Contents

Foreword

The great Danish physicist Niels Bohr famously (and humorously) pointed out that, 'Prediction is difficult, especially about the future.' While many thought leaders of our time have striven to improve on that situation, the inherent nonlinearity of all interesting systems on Earth keeps us thinking, probing and inventing the future—so as to have some hope of correct prediction. Prediction is even worse in the telecommunications and computing fields; a favorite quote of mine, published in the US magazine *Popular Mechanics* in 1949, opined that, 'Computers in the future may weigh no more than 1.5 tons.' True enough, it's quite hard to find computers that weight as much as 1.5 tons today!

Nevertheless, prediction is critically important, *especially* in the field of telecommunications as it collides, in slow motion, with the field of computing. A scant 40 years ago, the word 'network' meant something to do with telecommunications; now computing experts claim the word. That same 40 years ago, the new field of digital computing was irrelevant to telecommunications networks and services. Today all switching nodes are digital, all computers connect to a network, and confusion reigns as the very different qualities of service, pricing models and engineering reliability models cause those on both sides of the computing/telecoms divide to rethink what the service is and for whom it is provided.

It is unfortunate that our industry has a tendency to lump hundreds of technical leaps forward into neat and simple names like '4G,' but that's the world in which we live. You can't turn to a better resource than this book to find out what might be meant by '4G'—and why you need to understand the colliding telecoms/computing architecture, mobile and casually connected networks, semantics and ontologies, and intelligent agents. Someday the rock-steady 1950s-era telecommunications engineering will find its way into computing; and someday the intelligent, semantically aware systems of the 1970s will find their way into the network. This book prepares you to be up to date and aware when that happens.

The great American baseball player Yogi Berra also understood the difficulties of telling the future; he said, 'The future ain't what it used to be—but then again it never was!' We'll know more about the future when we get there, and it won't look much like we expected, but at least we can understand the foundational technologies that will make it work, so we can *invent* the future in time for our arrival. In the meantime, this tome will focus your thoughts on how those new foundations will support the delivery of innovative services on 4G networks and beyond, which will enhance all of our lives.

Enjoy this comprehensive overview of the foundations of the systems you will be building tomorrow!

Dr Richard Mark Soley
Chairman and Chief Executive Officer
Object Management Group, Inc.

Foreword

Congratulations on acquiring this volume about the latest advances and visionary insights in service provision. This book presents challenges and opportunities for future generations of mobile service delivery, pointing *Towards 4G Technologies*. It reaches out for an unparalleled way of networking and service provision.

Based on the vision of a ubiquitous service and networking environment, the authors explain how new concepts of mobile communication networks support an unprecedented mobile service environment. This new environment will be the open platform upon which new service concepts are built. These include methods for preference-based service discovery and selection, deeply personalized and mobile peer-to-peer applications as well as semantically enhanced and contextual intelligent services. Those principles overtake existing ones in unparalleled dimensions leading towards situation-aware and adaptable services. You will find this book structured in three main parts: the pillars of a new architecture, the foundations of smart service provisioning and the intelligent embedding of services into their environment. Thus it covers the complete range needed for successful service delivery.

The technology advances described in this book represent a rich set of innovations, but the authors do not loose sight of the fact that the user's needs have to be in the foreground of all considerations. Only then can successful business cases be constructed, allowing for a multitude of parties and players to find their business roles within the Mobile Information Society at large. Service provision continues to adapt efficiently on a global scale such that customers are able to invoke their familiar services wherever they journey to. *Services with Initiative,* that is proactively provided services based on the user's preferences and behavior patterns, are expected to become the new 'favorites', pointing towards easy and convenient use and a way out of information overflow.

Many of the underlying technicalities are discussed in this book in detail, from optimization of large system parts, to ontology design using the Semantic Web to represent services, to context-oriented programming, just to mention a few. It is my belief that the profound knowledge base represented in this book is beneficial to telecoms engineers, members of R&D departments, telecoms managers as well as academic researchers in computer science, electrical and electronic engineering and telecommunications. It provides the reader with a comprehensive understanding of next generation mobile service provision *Towards 4G Technologies* allowing for *Services with Initiative*

Please enjoy the following inspiring book chapters.

Dr Atsushi Murase
Managing Director of Research Laboratories, NTT DoCoMo

Preface

Publishing a book dedicated to advance service provisioning which includes the majority of related technologies appears to be both urgently needed and very challenging. Urgently needed, because only new and unprecedented service portfolios are capable of creating new revenue streams, something that all the involved business parties are looking for to resist heightening competitive market pressures. Challenging, because everyone who looks at this book has something between a distinct feeling and a vague idea of how services should look in the future and which capabilities they should possess to satisfy the user's needs. The aim of this book is to describe the various aspects of future service delivery and to add space for situational intelligence at service selection, invocation and during service run-time. It outlines the road ahead that promises to lead to a new service universe. The contributors' experience along the way is their Mobile Adventure. A new mobile service generation is destined to assist users in their daily life routine, with very flexible tasks and based on resources available to the user. New services need to understand how to adapt to user's environmental changes and act proactively in a way that the user feels guided and safeguarded on their way through the world of services. There are many building blocks required to complete the picture of 4G technologies, amongst them an underlying diversity of connectivity and an open programmable architecture that allows for system reconfiguration and optimization. The respective components are described in this book. Furthermore, advanced service support platforms that adhere to next generation service architectures and extend to a ubiquitous service environment are outlined. Taken together, the following book chapters respond to a new service provisioning paradigm which asks for Services with Initiative rather than cumbersome service discovery, non-optimized service selection, very limited user guidance and inconvenience during service execution.

What is unique about this book?

This book provides the first comprehensive overview of technologies that are needed for future networking and service delivery within next generation mobile systems. It consists of a thorough explanation of how personalization, peer-to-peer solutions, semantic computing, ontology engineering and description logic systems fit into this new concept. It explains why they will become a necessity for future mobile services. *Towards 4G Technologies* presents the latest challenges and opportunities for contextual intelligence, explaining the potential of context-oriented programming and a user-centric view on services and context. It covers hot topics such as intelligent user profiling, proactive service discovery and service selection. This book introduces seemingly diverse technologies and shows how they will play together to create a new more satisfying user experience with Services with Initiative as the enabler of a new mobile life style.

Who should read this book?

This book is set to become a helpful guide for evaluating the potential of future service opportunities for decision makers and business strategists, CTOs and telecoms managers alike. It is intended to provide telecommunications engineers in research and development departments, as well as academic researchers in computer science, electrical and electronic engineering and telecommunications with a comprehensive understanding of next generation mobile system technologies and services. For graduate students it provides invaluable reading to enhance their studies and brings useful insights on an all-embracing service support system view.

Structure and organization

Guided by forewords and the preface, the book opens the stage for the user's needs and expectations in future service provisioning. The introduction touches upon various aspects of service support mechanisms such as context awareness and personalization. It describes the problem scope as a complex service landscape and invites the reader to discover it through a 'Mobile Adventure'.

The three main parts are organized as follows: Part I explains the pillars that are needed for a new service support architecture. It starts with structuring mobile communication networks and takes the diversity of connection technologies and network topologies into account. The chapter on mobile communication networks is followed by an exhaustive description of service platforms, reviewing their functionalities by focusing on mobile service support. The next generation service architecture and its extensions towards ubiquity follow next. Mobile Peer-to-Peer system description and considerations of mobile middleware concepts are described afterwards. Part I concludes with system optimization methodologies based on Cross Layer Design. Part II elaborates on the foundations of smart service provisioning. The benefits of applying ontology technologies and description logics to increasing usability of semantic services are put at the center of considerations. This second part concludes with aspects of dynamic adaptation techniques of system parts and entire systems, thus describing living and growing components that form an open adaptive software foundation. Part III chooses a viewpoint that emphasizes the intelligent embedding of services in their environment. Starting with context-aware mobility management it leads to the description of context-oriented programming principles, proactive service discovery and selection mechanism and intelligent service composition. Part III concludes with a visionary view of enablers of modern mobile lifestyle, thus moving from Personal Mobility to Mobile Personality. The book concludes with an outlook that embraces the real world with Services with Initiative.

Hendrik Berndt
Munich, Germany

Acknowledgment

This book's inspiration is various research insights obtained over the years within the Future Networking Laboratory at DoCoMo Communications Laboratories Europe GmbH, Munich, Germany. The ideas and thoughts of many of the current and former colleagues from this laboratory have been influential in shaping this book. Naturally, their ideas are included throughout the chapters which follow. Moreover research associates we partnered with, most notably from DoCoMo Research Laboratories in Yokosuka, Japan, have given valuable advice on the research directions we took and from which we selected potential subjects for next generation mobile service provisioning as described here. Ideas from various collaborative efforts were gathered from working with universities and research institutions throughout Europe and inspired the contributors while writing this book.

The contributors' thanks go to all of the above mentioned sources of inspiration. Finally, I wish to thank the staff at Wiley for their professional support during the whole period of writing and the production phase. Without their continuous encouragement the publication of this book would be hard to imagine.

Hendrik Berndt
Munich, Germany

About the Editor

Hendrik Berndt holds the position of Chief Technology Officer and Senior Vice-President at DoCoMo Communications Laboratories Europe and is Director of the Smart and Secure Services and the Ubiquitous Networking research groups. In this capacity he is supporting DoCoMo Euro-Laboratories' research activities towards the 4th generation of mobile systems. Previously Chief Technology Officer of the Telecommunication Information Networking Architecture Consortium, headquartered in Tinton Falls, New Jersey, USA, he was chairman of the TINA Architecture Board and responsible for guiding technology developments and products in the area of Multimedia Services, Distributed Processing Environment and IP Control and Management. Until then he held several scientific and managerial positions in the Telecommunication Industry in Germany, was Executive Director of Advanced Technology for Global One in Reston, Virginia, USA and an invited member of Sprint's Office of Network and Architecture Planning in Kansas, USA.

Currently he is elected Steering board member of the European *e*-Mobility Technology platform initiative supported by the European Commission, Brussels, Belgium, member of the Board of Directors in the Object Management Group, headquartered in Needham, Massachusetts, USA and member of the curatorship of Fraunhofer Institute for Open Communication Systems, Berlin, Germany. He holds a Masters degree and a PhD in Electrical Engineering and was appointed from 2000 on as Visiting Professor at the Global Information and Telecommunication Institute, Waseda University, Tokyo, Japan.

About the Contributors

All contributors have been members of the Future Networking Laboratory at DoCoMo Communications Laboratories Europe GmbH, Munich, Germany. They possess an outstanding track record of publications and granted patents and have frequently been appointed to numerous technical committees at important international conferences on service provisioning, networking and software technologies.

Christian Bettstetter is Professor for Mobile Systems and Head of the Institute of Networked and Embedded Systems at the University of Klagenfurt, Austria. His research interests include networking, algorithms and protocols and modeling aspects of wireless networks. For eight years, his main research has been on multihop networking. Prior to becoming a professor, he was a senior researcher within the Future Networking Lab at DoCoMo Communications Laboratories Europe.

Zoran Despotovic is Senior Researcher within the Ubiquitous Networking Research Group at DoCoMo Communications Laboratories Europe. His research focus is distributed systems, peer-to-peer systems in particular. He holds a Dipl.-Ing. degree in Electrical Engineering from Belgrade University, Serbia and a PhD in computer science from the Swiss Federal Institute of Technology (EPFL-Lausanne, Switzerland).

Robert Hirschfeld is Professor of Computer Science at the Hasso-Plattner-Institut at the University of Potsdam. There he founded and leads the Software Architecture Group, developing methods and tools for improving the comprehension and design of large complex systems. He was a senior researcher with DoCoMo Communications Laboratories Europe where he worked on infrastructure components for next generation mobile communication systems with a focus on dynamic service adaptation and context-oriented programming. Robert Hirschfeld received a PhD in Computer Science from the Technical University of Ilmenau, Germany.

Wolfgang Kellerer is Senior Manager of the Ubiquitous Networking Research Group of DoCoMo Communications Laboratories Europe. His research interests include mobile service platforms, peer-to-peer, sensor networks and cross-layer optimization. He received his Dipl.-Ing. (MSc) and Dr.-Ing. degrees from the Technische Universität München, Munich, Germany in 1995 and 2002 respectively.

Marko Luther is Senior Researcher in Smart and Secure Services Research at DoCoMo Communications Laboratories Europe. He is an expert in formal methods, ontology technologies and context-aware mobile applications. He holds a Masters degree and a PhD in Computer Science.

Chie Noda is Assistant Manager of the Communication Device Development Department in NTT DoCoMo in Japan. She has been involved in telecom standardization activities and European research projects. She was Senior Researcher within the Future Networking Laboratory at DoCoMo Communications Laboratories Europe from 2001 until 2006. She holds a Masters degree in Mathematics and Science.

Massimo Paolucci is Senior Researcher in the Smart and Secure Services Research Group at DoCoMo Communications Laboratories Europe. He is an expert in semantic representation of services. His current work concerns near field communication based mobile applications. He holds a Masters degree in Computer Science from the University of Milan.

Christian Prehofer is Research Team Leader at the Nokia Research Center. His research interests are self-organized and ubiquitous systems, as well as software architecture and software technologies for mobile communication systems. He was Project Manager within the Future Networking Laboratory at DoCoMo Communications Laboratories Europe from 2002 until 2006. He obtained his PhD and habilitation in Computer Science from the TU Munich in 1995 and 2000.

Marco Sgroi is Manager in the Wireless Sensor Networks Laboratory sponsored by Pirelli and Telecom Italia, at Berkeley, California, United States. His research concerns platform-based design methodologies for communication networks and new application scenarios for sensor networks. He obtained a PhD in Electrical Engineering and Computer Sciences from the University of California at Berkeley in 2002. From 2004 until 2005 he was Senior Researcher within the Future Networking Laboratory at DoCoMo Communications Laboratories Europe.

Matthias Wagner is Senior Manager of the Smart and Secure Services Research Group at DoCoMo Communications Laboratories Europe. He has specialized in proactive services, context awareness and concepts to leverage the semantic web in mobile applications. He holds a Masters degree and a PhD in Computer Science.

Jörg Widmer is Project Manager of the Ubiquitous Networking Research Group at DoCoMo Communications Laboratories Europe. His research is concerned with MAC layer design, network coding, algorithms for wireless multi-hop networks and future networking architectures. He obtained his Masters degree and PhD in Computer Science from the University of Mannheim, Germany in 2000 and 2003, respectively.

1

Introducing the 4G Mobile Adventure

Hendrik Berndt

1.1 The User Takes Centre Stage

Users don't care about technologies; they want services that fulfil their needs and they care about prices. As interesting and exciting technology advances develop over time, the real mobile adventure for users consists of discovering new services and running them with ease. Future services should be provided in a sophisticated and unobtrusive manner, cheap enough for customers to use them without thinking too much about the cost. Service offerings should be easy to find. Intelligent 'helpers' would be available that can be asked to guide the user into a life-unburdening and entertaining service space, creating a feeling that every task at hand becomes a little easier to carry out. Information overflow should no longer bother customers, as is too often customers' complaint nowadays. In such a service environment the user will be delighted to take centre stage.

As a starting point for discussion, Figure 1-1 provides a simplified view of the evolution of mobile communication systems from the 1st to the 4th generation, highlighting their respective major characteristic. The figure shows how the generations are mainly distinguished by significant technology enhancements. Each of the previous generations has had a lifespan of nearly a decade. This time period includes the research and development time necessary for specification of the technology, as well as its development and deployment for commercial use. Even though today wireless applications and services are already having immense business success and include Information Data Services, Business Services (for example, Mobile Banking and Mobile Payment), Entertainment Services (for example, Karaoke, Online Games and Streaming Audio/Video), Enterprise Services (for example, Office Document Applications and Intranet Applications), as well as Communications Services (for example, traditional Voice, SMS and email), it is in future systems, beyond the third

Towards 4G Technologies. Edited by Hendrik Berndt.
© 2008 John Wiley & Sons, Ltd.

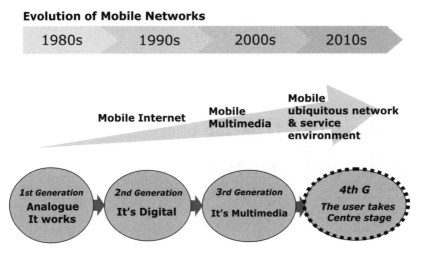

Figure 1-1.   Evolution of wireless communications by decade

generation (3G), that the user will finally take the driver's seat and manoeuvre through an ever-growing service universe.

Where precisely do we stand at the time of writing this book?

The successful early adoption of 3G IMT-2000 mobile services on a global scale, providing both traditional voice and high-speed data services, such as email and instant messaging, multimedia and Internet access, goes in line with the convergence of traditional mobile and broadband. The interest in wireless broadband today already indicates its potential mass-market appeal.

We take as an example NTT DoCoMo, one of the world's leading mobile communications companies. 'As of September 2007, DoCoMo serves 40 million customers, who have sub-scribed to the world's first 3G mobile service, launched as FOMA™ in 2001. DoCoMo also offers a wide variety of leading-edge mobile multimedia services, including i-mode™, the world's most popular mobile email/Internet service, used by more than 47 million people in Japan alone. With the addition of credit-card and other e-wallet functions, DoCoMo mobile phones have become highly versatile tools for daily life' (see <http://www.nttdocomo.com>).

Following the logic of Figure 1-1, global research efforts for beyond 3G systems should had taken off in the early 2000s. And indeed they did, pointing towards emerging solutions described as Super 3G, 3G Long-Term Evolution, IMT Advanced and 4G. Related work is most notably concerned with a new air interface. The most promising interface seems to be based on multi-carrier transmission, providing data rates of 100 Mbit/s over wide areas and up to 1Gbit/s in hot spots. It goes without saying that, as a necessary prerequisite for system deployment, decisions have to be made about allocating early enough the frequency bands needed for next-generation mobile systems.

As far as services are concerned, however, the road towards 4G leads to unprecedented perspectives. In the future, it will not be the single service that will satisfy the user and create a loyal customer; rather it will be the embedding of the desired service in a suite providing user support for service discovery and runtime conditions that provide guidance through

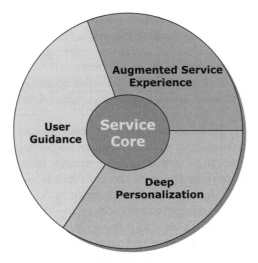

Figure 1-2. Universal Service Provisioning Model

service usage. Actively selecting the service according to the user's preferences and its adaptation through a context-aware service delivery mechanism will become a commodity. When these ideas materialize, a new 'service universe' will have been created.

The Universal Service Provisioning Model, a new generation of the TINA USCM,[1] provides such enhanced principles for future service delivery, as shown in Figure 1-2. The core represents the principal characteristics of the service, for which it was built. After adaptation it will be delivered as personalized instantiation of the desired service, created through awareness of the ambient situation. Simultaneously, the service will be accompanied by a guidance mechanism offering the user an easy understanding of the service. The augmented service experience could possibly be enhanced through elements of virtual reality. Human senses that, in the past, there have been no attempt to communicate can become part of this new Service Universe. Further extensions will allow for services that are capable of proactively engaging in their invocation and execution. These are explained and labelled throughout the book as 'Services with Initiative'.

The aim of this book is to provide the necessary pillars of a new architecture and prepare the stage for a new suite of services that are – compared with today's service offerings– semantically enhanced and destined to be proactive if the customer's situation and preference allows for it, aiming to provide a higher level of user satisfaction.

Ambient connectivity and the most advanced service delivery mechanism, supported by a sophisticated service provisioning platform and methods that allow for actions based upon contextual intelligence, are its most prominent components.

1.2 Business Model Considerations – the Need for an Open Programmable Architecture

The traditional business model for telecommunications services involved two parties only, arranged in a vertical 'closed' service model. In this Provider/Consumer model, the telecommunications operator acts in a business role that provides both the network infrastructure and the service. Over time more and more enhanced service features were included, but the core of the business model was an end-to-end closed and proprietary solution that locked service consumers to the operator for connectivity and services.

This closed vertical service model required the telecommunications operator to maintain all infrastructure and services and consequently often excluded the best-of-breed third-party services and solutions.

Today progress towards more openness has been made in many ways. Open APIs allow content and application providers to offer their services to customers, often enhanced with additional functionality such as integrated billing or security features, provided by the tele-communication operators. In particular, more and more mobile operators tend to adopt what is called a 'Semi-Walled Garden' business model, which allows users to obtain content and services from players officially recognized and endorsed by the mobile operator, but does not restrict the customer from invoking services from any other 3rd party service provider.

However, the future success of a new generation of mobile systems will rely on a new business model that evolves from the one described above in a way that will unlock new levels of openness and adaptability. The model should allow the participation of all newcomers to the business and should be capable of incorporating new business roles and refining the current role of operators, services and bandwidth brokers, application providers, subscribers and users, thus providing new benefits to all stakeholders.

Such a new business model needs to build upon an open, adaptive and programmable system architecture with the following properties:

- Its boundaries should be clearly defined through reference points and open APIs.
- Its parts should ensure interoperability between all parties, their respective business roles and the services provided.
- The different evolution cycles of the individual parts of the system are built for quick adaptive actions.

As in past designs of large mobile communication systems with many independent or interdependent components, we advocate a layered architectural approach. The use of layers as the architectural pattern is based on a vertical structuring, like a layered cake, where the stack of vertical layers is considered orthogonal to each horizontal operational layer. Thus, the layering pattern supports vertical information being passed from higher-level layers to lower-level layers and vice versa.

4G systems are expected to adjust efficiently within a fast-changing environment from radio transmission to applications and thus require extended evolutionary and adaptation capabilities. The goal is to provide dynamic adaptation mechanisms that are capable of handling the level of complexity that is forced upon us, stemming from concerns about multiple users, devices, platforms, services and contexts. Adaptation should therefore provide support for the ability of application-layer input to react to changes based on information (profiles, preferences, context information) provided through personalization and context awareness. Adaptation should also be capable of reacting on the basis of lower-level protocol information such as available data rates or current channel conditions. Typical situations for adaptation include a substantial change in connectivity characteristics, the user entering a new service domain and a change of terminal device during the service session. Accordingly, we can separate adaptation into:

- media adaptation, such as text-to-speech or transcoding
- content adaptation, such as adapting the presentation or re-ordering information
- service behaviour adaptation affecting the service logic

Adaptation must be both reactive and proactive. This means adaptation can be requested by the user directly, for example the transformation of text to speech, or it can be performed automatically based on context information, for example adapting the presentation of information to the limited display of a mobile phone.

So as to achieve a generic solution for adaptation, we advocate an architecture that consists of abstraction layers, each of which is composed of one or more open platforms. Thus, not only open interfaces between layers are supported. Figure 1-3 depicts the open platform concept,[2] where the platform is built of a platform base and platform components that can be modified to address different needs based on situational requirements.

An example of part of an open programmable architecture with cross-layer collaboration capabilities is depicted in Figure 1-4 as service and middleware platform, where components and interfaces that are subject to adjustment form an adaptation unit.

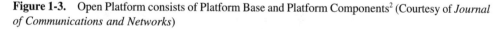

Figure 1-3. Open Platform consists of Platform Base and Platform Components[2] (Courtesy of *Journal of Communications and Networks*)

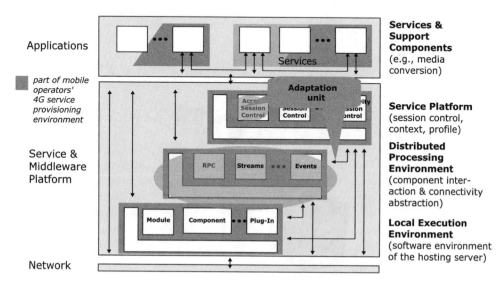
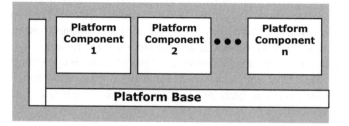

Figure 1-4. Configurable service and middleware architecture enabled for cross-layer collaboration

Such a structured open programmable architecture (see Section 2.5) with its built-in cross-layer collaboration would easily reduce the efforts required to introduce new services and new service support features based on existing systems. With the concepts of adaptation units and variation points – points that are subject to efficient revision of software components – it allows flexibility and adaptability to rise to a new level so as to be able to handle unanticipated system evolution, such as may be required by 4G mobile communication systems. With such enhanced ubiquitous computing platforms, context-dependent service adjustments and auto-adaptive services – services that can act on their own initiative – will become reality.

1.3 Ubiquitous Service and Network Environment

Throughout their history computers have become smaller and cheaper as a result of continuous advances in digital electronics. At the same time, processing power and storage capacity have grown at an amazing rate. The combination of these two trends has allowed the introduction of computational capabilities into places and devices that would have previously been impossible in the earlier state of technology. In addition, it can be observed that more and more everyday items possess embedded wireless interfaces, as depicted in Figure 1-5, allowing networking amongst them. This opens the door to new and interesting business opportunities, where everyday life support, healthcare, robotics and logistics are just a few of the possible application areas. In the near future people will be surrounded by ever more computers, leading to the situation in which we are 'dwelling' or living amongst them. Such a scenario is described as a ubiquitous service and networking environment, an execution environment in which over time users will gradually be surrounded by computational resources and tiny networked devices.

Figure 1-5.   Everyday items will soon have their own wireless interfaces

New challenges arise from such an environment and are described in the chapters of this book. In this introduction only a few are referred to: first, the fact that a myriad of elements need to be dynamically configured in order to build a meaningful – often application-dependent – network topology. These elements need to be controlled and maintained to make it possible to run new services and applications upon them. To manage such a ubiquitous environment, centralized control methods are rather limited. Nature provides a more favourable solution in which self-organization is key to developing complex global procedures from simple local rules. Some basic principles of self-organized systems are depicted in Figure 1-6.

An extended self-x behaviour, where x stands for organization, healing, configuration etc., allows for building adaptive, universally accessible, easily configurable and fair (in terms of usability) networks, which together produce a complexity that is far beyond the capabilities of each single element. Robustness is also supported by the large number of nodes, so that, even if single nodes fail, reliability is still protected.

The second challenge arises because the potential of networked elements varies considerably, from, for example, (active) RFID tags, battery-powered sensors and short-lived devices to devices with a permanent power supply. A fairly generic characteristic of a ubiquitous networking environment is the fact that elements are often power-constrained. This means that long links to base stations are not always feasible so that multi-/hop-access scenarios become necessary and incentives for networked elements to act as relays have to be introduced.

The third challenge is that, unlike many other communications systems in a ubiquitous service and networking environment, the networking and the applications are highly interdependent, and often integrated. Sensors are typically designed for only one simple application. Today, methods and technologies that extend usability and overcome shortcomings are used to 'fine tune' the purpose and aim of a particular network, or example through a 'mode

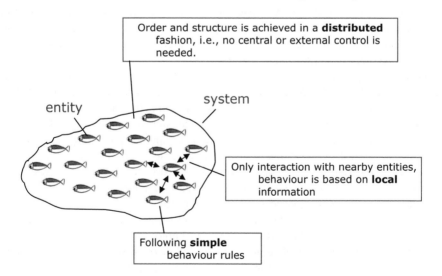

Figure 1-6. The benefits of self-organization: the desired system behaviour emerges solely from simple local interactions with nearby entities

switching' mechanism, where sensor networks react upon the changing environment by changing network topologies.

Fourth, despite the multitude of networked technologies and the variety of services, seamless access in the next generation of mobile communication systems has to be granted to the user. A wide range of access technologies and access networks with minimal input from the user will go far beyond the 'roaming' as we know it today (see Section 2.3.1.) For service provision, seamless access means allowing the users to reach their personal services and profiles anywhere, at any time and over all access networks and devices. This requires advanced mobility concepts, not only in the network layer, but also in the application layer of a mobile system. We will discuss the different mobility paradigms such as terminal mobility, personal mobility, service mobility and session mobility in more detail in Section 3.3.1.

Last, but not least, security and privacy concerns turn out to be of prime importance in a scenario where everyone and everything is connected. As a result of their extended capabilities, personal devices operated in an ubiquitous service and networking environment are able to provide authentication and authorization not only for their own usage, but also for other kinds of mobile and non-mobile services, utilizing, for example, near-field communication methods to invoke services in order to obtain e.g. train tickets or to provide corporate access, etc.

There are many approaches to introducing measures to counter security threats and privacy violations. One notable example solution for neutralizing malicious node behaviour in a ubiquitous environment can be provided by mobile-operator-mediated communication. In such a scenario provider-assisted devices such as mobile phones or personal digital assistants become key components in enhancing the capabilities of, for example, ad hoc networks or other ubiquitously networked elements. Those devices that had previously established a customer–provider relationship with any mobile operator can be seen as an overlay network of trusted nodes, as depicted in Figure 1-7.

These provider-assisted nodes exchange routing information over several paths in order to detect information anomalies. Consequently, they have more efficient means to detect malicious nodes and possess rerouting capabilities based on their better knowledge of the network as a whole; this is illustrated in Figure 1-8.

Summarizing this introductory section on the ubiquitous service and networking environment, we emphasize that future mobile system generation will have to meet the challenges of extending connectivity and service provisioning on an unprecedented scale. Involving both human-to-human and machine-to-machine communication, there will be no barrier to the inclusion of artefacts and living things into a user's all-embracing environment. We believe that it is in the interest of stakeholders such as mobile operators' to support new business opportunities by integrating ubiquitous and cellular network structures. That is why we conclude this section with a convincing application scenario that takes into account the fact that some of us forget where we placed our belongings. The scenario takes advantage of the opportunity to tag everyday objects that are valuable to the customer and, as such, made 'sense-able' through object sensors that use radio technologies such as RFID readers, Bluetooth or similar technologies. Mobile-network-enabled devices would now act as a 'ubiquitous gateway' between the cellular infrastructure and the extended ad-hoc sensory network environment. Queries about the location of a lost object could be sent by a scoping query manager to a set of remote ubiquitous gateways that trigger a response whenever the lost

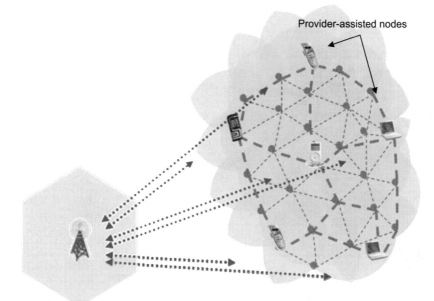

Figure 1-7. Provider-mediated communication

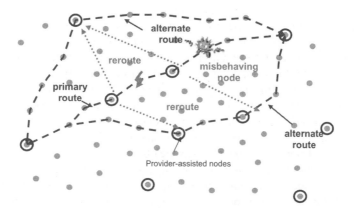

Figure 1-8. Detection of malicious nodes and establishment of an alternative route through provider-mediated overlay nodes

item comes within sensing range or when a message is transmitted through a nearby object sensor node. Figure 1-9 depicts this particular application scenario.

1.4 Context Awareness

A great opportunity in leveraging the real power of mobile applications as Services with Initiative is to aim to provide the ability to detect and react to environmental variables at

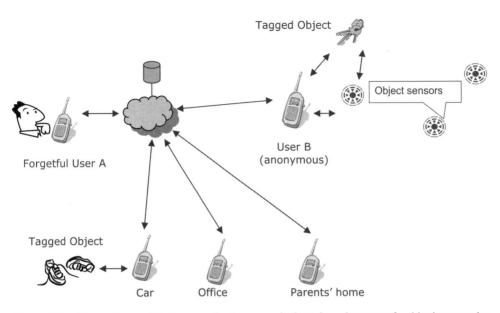

Figure 1-9. 'Remember and Find', an application scenario that takes advantage of a ubiquitous service and networking environment and finds misplaced, lost or stolen personal items via mobile networks

service provisioning. Environmental variables – usually described as the context – consist of accessible information about the user's situation at service runtime. The assessment of the user's situation can vary from the interpretation of a single contextual data item, such as location, to composed inputs based on multiple data items. Context categories include, but are not restricted to: computing context, for example network connectivity; time context, such as the time of the day; physical context, such as the lighting level; the relationship between the application and nearby objects; and even the user's social situation. The context also may include aspects such as interactions previously performed between the user and the service (application history) and future expected relationships. Figure 1-10 depicts the situation where different contextual inputs are observed and combined to create a specific context space, relevant to a particular user or user group at the time of service provisioning. Context-Aware Service Provisioning, based on Context Spaces, allows the establishment of a framework for the creation of scoped contextual boundaries and the sharing, storage and selection of context entry objects within those boundaries. However, even the most complex context space scenario only represents a partial view of the full set of situational variables.

For example, a user can move with an application from one situational context to another and in the new context be challenged with even more contextual heterogeneity (Figure 1-11). Moreover, further environmental properties representing dynamicity, continuity and non-determinism increase the complexity of grasping the environment. The state of the environment will change dynamically, independently of the actions of the service. There is no discrete fixed number of states forming a context. Furthermore, there is no guarantee that the available state information is accurate and the service will have only partial control of the state of the environment. Within such a diverse understanding of the context, Services with

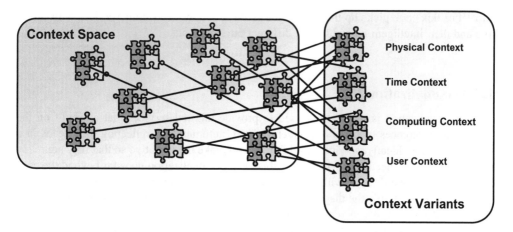

Figure 1-10. Context-space concept building upon different context variants

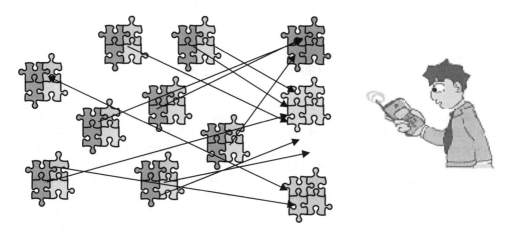

Figure 1-11. The user and his service embedding in a complex contextual environment

Initiative have to react within a highly dynamic scenario that is unlike any one that has existed previously.

To handle the environment described above one needs efficient ways of:

- coherent and distributed context modelling
- context search, information gathering and abstraction
- context grouping and filtering

In addition, the following will be needed:

- adaptive algorithms capable of gathering knowledge from samples
- ambient reasoning components to interpret the situation at hand
- generalization mechanism to apply that knowledge to new situations

Part III of this book picks up the challenges given above by describing Services with Initiative and their intelligent embedding in different environment.

1.5 Personalization

Deep personalization is the method used to provide tailored services that are built on the individual preferences of users in a given context, automatically reflecting user needs in a specific situation. Ideally, services would proactively take the initiative so that they are ready for use when really needed. By its nature, this deep personalization goes far beyond the configuration of mobile devices with personal data or a simple adaptation of content to device capabilities, for example the display size. Personalization targets a complete user model that can be used for all tasks in service provisioning. Recent standardization efforts concerned with user identity and personal access only include basic user modelling and profiling. These include:

- CC/PP (World Wide Web Consortium Composite Capabilities/Preferences Profile – http://www.w3c.org)
- NET PASSPORT (http://www.passport.net) profiling as discussed within the Liberty Alliance Project (http://www.projectliberty.org)
- user profiles proposed in the 3GPP (http://www.3gpp.org) and the Parlay Group (http://www.parlay.org)

Beyond user modeling and profiling there are a number of further challenges for the service and network infrastructure involved in personalization. Users will no longer have only one single communications device, but use several different devices concurrently; some of these, such as public Internet terminals, they will acquire from the environment. Thus, the availability, i.e. the download, of profile information and the synchronization of changes will be necessary. Appropriate signalling mechanisms and discovery technologies will have to ensure that the right information is provided at the right place at th right time. However, the exchange of user profile information raises security and privacy concerns. All service providers have to make sure that no personal user data or data about the user's context is disclosed (confidentiality) or modified (integrity). The trade-off between personalized services and the disclosure of personal information is a critical issue for service provisioning and customer retention. The latter is very much related to trust, which is not only limited to the security context, including, for example, confidence on the part of the customers that the service provider is always acting in their interest. While trust is very much based on security and privacy, it is also concerned with reputation.

So far we have looked at personalized service provisioning from the viewpoint of a user invoking only a single service. However, the ultimate goal in personalized service provisioning has to be the fulfilment of individual user needs, often expressed as complex tasks that are typically further divided into simpler sub-goals and subsequently matched to different services. If we take the example of a business traveller, it can be shown how ubiquitous services may accompany a user throughout his daily travel routine and how related services like renting cars, finding the next ATM machine and booking a convenient hotel, have to work as a unit towards the user. Such services – especially when accessed via mobile devices

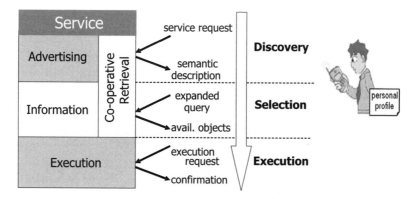

Figure 1-12.   Personalized service discovery, selection and provisioning

– demand a more thorough support for personalization that cannot yet be achieved by today's techniques.

With the number and diversity of services expected to grow, adequate techniques for user-centric and preference-based service discovery and selection will be needed. Even though UDDI and WSDL are commonly used today to implement service catalogues, they still essentially lack strong concepts for service personalization, which are crucial for advanced usability. To solve this problem, service cataloguing should be enhanced with concepts from cooperative databases and collaborative filtering to discover, select and combine services according to the special needs and preferences of the individual user. Our view of personalized services is shown in Figure 1-12. User information – such as profile data and user preferences – is crucial for all the phases of service provisioning, namely: service advertising, service discovery, service selection; and service execution.

While we deal with personalized selection and execution later in the book, in this introduction we only briefly describe the early steps in the service provisioning of personalized advertising and discovery of services according to the special needs and preferences of an individual user. Typically usage patterns and enriching service requests, with the explicitly or implicitly given preferences of single users or user groups, can be used to create essentially improved user–service matching.

1.6 Mobile Adventure

The mobile adventure begins now that we pursue our outlined vision of deeply personalized, situation-aware and proactive service provisioning as we bring forward the detailed challenges in the following chapters. This adventure is meant, along with all technical particularities, to add a human touch of understanding of a user's intentions and needs to the field of communications business where information overflow is the rule rather than the exception. We want to introduce the building blocks that, joined together, fill the vision and shape a nearly complete picture. Figure 1-13 illustrates our mobile adventure service playground.

The adventure builds upon a ubiquitous networking base that includes a diverse set of types of connectivity and communication channels, overlay and underlay networks, with wireless communication obviously being the most challenging one. It contains relay-enhanced cells,

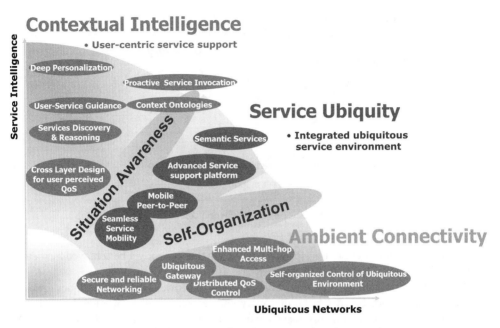

Figure 1-13.   The mobile adventure service playground

multi-hop access, ubiquitous gateway functionalities, it allows for quality of service assurance and secure and reliable networking. Connectivity naturally has to be controlled, managed and maintained as necessary in an efficient way. That's why new methods of organization are created, a prominent one of them being self-organization. Additional management functionalities can be provided by thin management overlays that are dynamically adaptable to the task at hand [3].

Service Support Platforms provide generic service features that are applicable to many services and application classes, and by running them, the platform provider can gather a deep understanding of the ways the platforms are used. As a trusted stakeholder, the service support platform provider might add components that deal with reputation management, build a trustworthy environment for customer relations and apply flexible accounting, charging and billing solutions. Service support platforms provide the means for seamless service session mobility, independent of the underlying connectivity and can render a familiar touch and feel of services, known to the user. Semantically enhanced services can introduce and represent themselves in a more sophisticated manner, resulting in a user-friendly service environment.

Finally, in our vision contextual and situation-aware intelligence will be applied in each and every step of service discovery, invocation and execution. This intelligence is aiming at service delivery that takes into account the assets and resources of the user, his interests and preferences, the user's level of familiarity with the desired services and, if necessary, combines it with user guidance for the services provided.

Naturally further enhancements can be imagined, such as adding components of virtual reality, tactile senses etc. to the playground. We refer to those briefly in the outlook at the end of the book.

Part I

The Pillars of a New Architecture

2

Mobile Communication Networks

Christian Prehofer, Christian Bettstetter and Jörg Widmer

2.1 Wireless Technologies at a Glance

Today's mobile services are made available through a broad spectrum of modern technologies for wireless communication. The goal of this section is to give an overview of the state-of-the-art as well as a preview of forthcoming technologies, so that the reader will become familiar with the application scope and key features of these technologies. Our focus is on technologies that are standardized and widely deployed in products on the consumer market. In this context, the term 'wireless' means that the user's devices are not connected via cables to some network infrastructure or to other devices. In most cases, the wireless transmission is implemented using radio waves, but also light waves or other transmission media are possible.

2.1.1 Classification

As shown in Figure 2-1, we can classify wireless technologies according to their scope, that is, whether they are used to form a personal, local, metropolitan or wide-area network. The borders between these categories are not strict, but rather give a rough framework for categorization.

Personal area networks are networks between mobile devices carried around by persons or worn on their body. A typical example is the interconnection between a music player or mobile phone and the corresponding headphones or a head-mounted display (see Figure 2-2). If we look further into the future, personal area networks will interconnect different components of a wearable computer. Current wireless technologies for this purpose are mainly developed by the IEEE (Institute of Electrical and Electronics Engineers) in its 802.15 series of standards. These standards include the technologies ZigBee, Bluetooth and Ultra-Wide-Band (UWB). Alternative ways for wireless transmission in personal area networks (not shown) are infrared communication (for example, as covered by the IrDA standard) or emerg-

Towards 4G Technologies. Edited by Hendrik Berndt.
© 2008 John Wiley & Sons, Ltd.

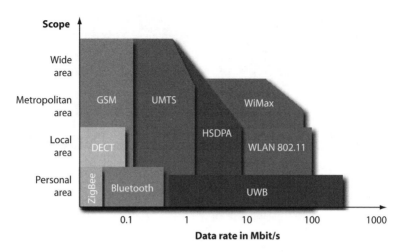

Figure 2-1.   Classification of wireless communication technologies

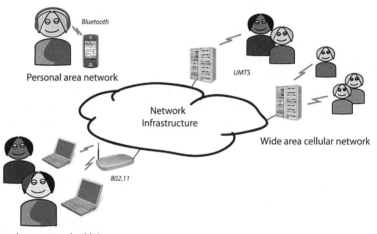

Figure 2-2.   Usage scenarios for wireless communication technologies

ing technologies for transmission over the skin (such as NTT RedTacton). The range of these technologies is of the order of some metres. Depending on the intended application, low to high data rates, up to more than 100 Mbit/s, are offered.

A local area network is a network between devices in a house or office. This includes, for example, a network between stationary and mobile computers, printers, servers, displays and television sets. Today's most popular examples of wireless local area networks (WLANs) are products based on the IEEE 802.11 standard. This technology – which is also called 'WiFi' – has been a success story over the past few years. So-called WLAN 'hot-spots' have been installed at universities, companies, airports, conference venues and other public places, to give visitors wireless access to the Internet and their emails. With the increasing penetration of broadband Internet access at home, private WLANs in people's apartments and houses are also becoming more common. The latest 802.11 products feature data rates of several tens of Mbit/s at a range of the order of some 100 m, depending on the environment. Another example of a local area network is the European DECT (Digital Enhanced Cordless Telecommunication) standard. It is used for cordless telephony in millions of households.

A metropolitan area network typically covers a city or metropolitan area. A typical example of a wireless metropolitan area network (WMAN) is the newly developed WiMax standard (IEEE 802.16). It offers high-rate Internet access at homes and offices with low user mobility. The typical transmission range of a WiMax transceiver is of the order of 10 km and above. In this way, it provides a broadband wireless solution for the 'last mile' between the network operator and the user in their home or office. It connects the operator network either to a locally installed WLAN or directly to the individual end devices. Its major advantage compared to wired access networks, such as fibre-optic and digital subscriber line (DSL) links, is the cheaper deployment cost. Another WMAN technology is IEEE 802.20.

A wide area network usually covers a complete country or even goes beyond national borders. It offers connectivity 'everywhere at any time'. The network is typically operated by a company or a government institution. Two prominent example of wide area networks for wireless communication are the Global System for Mobile Communication (GSM) and its successor the Universal Mobile Telecommunication System (UMTS). Users sign contracts with a network operator and can then use their mobile phones for telephony, text messaging and data applications. The data rates offered by UMTS are of the order of several 100 kbit/s. An extension to UMTS, High Speed Downlink Packet Access (HSDPA) offers packet-based data services with data rates up to 10 Mbit/s. Another example for a wide area wireless service is the Digital Audio and Video Broadcast (DAB and DVB) system. Satellite communication systems, which actually cover the whole world, also fall into this category.

Tables 2-1 to 2-3 summarize the key technical features of the mentioned technologies. In the following, we discuss some of them in more detail.

Table 2-1. Key technical features of wireless technologies: WPANs

TECHNOLOGY	ZIGBEE WITH 802.15.4	BLUETOOTH (802.15.1)	UWB
Typical usages	Networking of small devices and sensors	Cable replacement for head phones	Synchronization of multimedia data; wearable computing
Typical range	<50 m	10 or 100 m	4 or 10 m
Data rate	40 kbit/s, 250 kbit/s	max. 1 Mbit/s	>100 Mbit/s
Frequency	868/915 MHz, 2.4 GHz	2.4 GHz unlicensed	3.1–10.6 GHz
Physical layer	DSSS	FHSS	Impulse radio
MAC layer	CSMA/CA and slot allocation	TDMA	TDMA
Backwards compatibility	n.a.	n.a.	n.a.
Year of standard	2004 (2003: 802.15.4)	1998	Ongoing
Products since	2004	2000	2007

Table 2-2. Key technical features of wireless technologies: 802.11 WLANs

TECHNOLOGY	802.11	802.11b	802.11a	802.11g	802.11n
Typical Usages	Wireless access to the Internet	Wireless access to the Internet	Wireless access to the Internet	Wireless access to the Internet	Wireless access to the Internet
Typical range	~30 m indoors ~200 m outdoors	~30 m indoors ~200 m outdoors	~15 m indoors ~100 m outdoors	~30 m indoors ~200 m outdoors	~100 m indoors
Data rate	max. 2 Mbit/s	max. 11 Mbit/s	max. 54 Mbit/s	max. 54 Mbit/s	>100 Mbit/s
Frequency	2.4 GHz unlicensed	2.4 GHz unlicensed	5 GHz licensed and unlicensed	2.4 GHz unlicensed	2.4 GHz unlicensed
Physical layer	FHSS/DSSS	DSSS	OFDM	OFDM	OFDM with multiple antennas (MIMO)
MAC layer	CSMA/CA	CSMA/CA	CSMA/CA	CSMA/CA	CSMA/CA
Backwards compatibility	n.a.	802.11	none	802.11b	802.11b
Year of standard	1997	1999	1999	2003	Expected 2009
Products since	1998	1999	2002	2003	2006 (pre-standard products)

Table 2-3. Key technical features of wireless technologies: DECT, cellular networks, WiMax

TECHNOLOGY	DECT	GSM / GPRS	UMTS / HSDPA	WIMAX (802.16/802.16e)
Typical Usage	Cordless telephony at home	Mobile telephony, text messaging	Mobile telephony, multimedia messaging, Internet access	Wireless Internet access
Typical range	~50 m indoors	~1–5 km in cities, max. 30 km	Several 100 m in cities	~3–5 km
Data rate	32 kbit/s	Typical 30–70 kbit/s max. 170 kbit/s	Typical 200–500 kbit/s max. 2 Mbit/s max. 10 Mbit/s with HSDPA	max. 75 Mbit/s per channel
Frequency	1900 MHz	900 MHz 1800 MHz 1900 MHz licensed	2 GHz licensed	802.16: 10–66 GHz band (line of sight needed) 802.16a: 2–11 GHz band, both licensed and unlicensed
Multiple access	TDMA	TDMA	CDMA	10–66 GHz: TDMA with TDD and FDD 2–11 GHz: OFDM
Year of standard	1992	1990	1999	2001/2005
Products since	1993	1991	2002	2006

2.1.2 Wireless Local Area Networks: IEEE 802.11

The main idea behind WLANs is to extend the local area networking functionality of mobile computers to the wireless domain, in other words to replace the cable between a computer and the network by a wireless transmission. As illustrated in Figure 2-2, a mobile computer establishes a wireless communication link to an access point (AP), which connects the device to the infrastructure. In this way, WLANs provide very convenient and fast access to the Internet in public venues, companies and at home. Users no longer need an Ethernet cable and can move around freely within a certain area.

The working group 802.11 of the IEEE has defined a standard for wireless local area networking. It specifies the physical and medium access control (MAC) layer of the wireless link between a device and its AP or between two devices. The original IEEE 802.11 standard was published in 1997. Several enhancements were made in the following years to support higher data rates and additional functionality. Today, there is a whole family of 802.11 standards (see Table 2-2 and Figure 2-3). The following paragraphs highlight the main characteristics of these standards.

2.1.2.1 Physical Layer

The physical layer defines the wireless transmission between a mobile device and the AP or another device. The original 802.11 standard supports a data rate of up to 2 Mbit/s in the ISM (industrial, scientific and medical) band at 2.4 GHz. This frequency band can be used worldwide free of licence. Two radio techniques are defined: direct-sequence spread-spectrum (DSSS) and frequency-hopping spread-spectrum (FHSS). In DSSS each device multiplies the data by a pseudo-random chip sequence. Since the chip sequence has a much higher frequency than the data, the resulting signal has a much wider frequency bandwidth than the original data ('spread spectrum'). The receiver can decode the original data using the same pseudo-random sequence. In FHSS a similar approach is made by changing the carrier frequency rapidly.

The original 802.11 standard was soon replaced by 802.11b, which offers a maxmum data rate of 11 Mbit/s with a typical transmission range of 30 to 50 m indoors and about 100 m outdoors. It works in the same frequency band as 802.11, uses DSSS and is backwards compatible. The standardization of 802.11b led to a widespread use of WLAN technology in mobile devices.

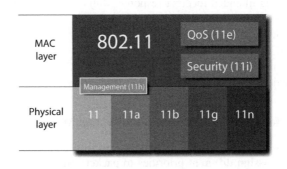

Figure 2-3. Overview of 802.11 standards

Higher data rates are achieved with 802.11a, offering a maximum data rate of 54 Mbit/s using Orthogonal Frequency Division Multiplexing (OFDM). OFDM divides the data stream into many sub-streams and transmits them simultaneously over different frequency channels. One of the main advantages is the good resilience against frequency-selective fading. In contrast to 802.11b, 802.11a transmits in the 5 GHz band and is thus not backward compatible to 802.11b, unless the equipment implements both standards. The higher carrier frequency has advantages and disadvantages. On the one hand, the 5 GHz band is used by fewer technologies than 2.4 GHz, resulting in lower interference. On the other hand, the high frequency reduces the range and limits the transmission to almost line of sight.

The 802.11g standard combines the higher data rate of 802.11a with backward compatibility to IEEE 802.11b. It operates in the 2.4 GHz band and offers data rates up to 54 Mbit/s. The standard defines four physical layers: DSSS (as in 802.11b), OFDM, an enhanced DSSS and a combination of DSSS and OFDM.

The latest WLAN standard is 802.11n. It achieves 100 Mbit/s and has a significantly increased transmission range. It does so by using multiple antennas. Although 802.11n is still under development, so-called 'pre-standard products' are already available on the market.

2.1.2.2 Medium Access Control (MAC) Layer

The MAC layer controls the access of different devices to the shared wireless medium. IEEE 802.11 specifies a distributed coordination function (DCF) in which all devices compete for medium access following a CSMA/CA (carrier sense multiple access with collision avoidance) protocol.

As the name says, the protocol employs the concept of carrier sensing: If a device intends to transmit a packet, it first listens to the medium to ensure that no other device is transmitting. If the medium is free for a certain period of time, the device is allowed to transmit. If the medium is busy, however, the device defers its transmission to a later point in time. It sets a random backoff counter that is decremented during periods in which the channel is free. When the backoff counter reaches zero, the device is allowed to transmit.

When this mechanism is applied, there is still a non-zero probability that two or more devices transmit at the same time, thus causing packet collisions. Hence, an optional handshake mechanism has been introduced. It works as follows. If a device wants to send a payload packet, it first has to send a short ready-to-send (RTS) packet, which contains the length of the payload packet. The intended receiver responds with a short clear-to-send (CTS) packet if it does not have an ongoing reception from another device. After this handshake succeeds, the transmitting device can send its payload packet. Finally, the receiver responds with an acknowledgment (ACK) packet.

In addition to communication to an AP, the 802.11 MAC layer also supports an ad-hoc mode. It enables the devices to communicate directly with one another without needing an AP.

2.1.2.3 Extensions for Quality of Service, Security, and Management

The 802.11 suite also defines a Quality of Service (QoS) mechanism. The 802.11e standard allows the devices to assign different priorities to packets on the MAC layer. Packets with high priority gain access to the medium faster than packets with low priority. Four QoS categories are defined: voice, video, best effort and background traffic. Voice has the highest and background traffic the lowest priority.

Another extension has been made to make 802.11 networks more secure: The 802.11i standard includes enhanced encryption and authentication methods, such as key management and secure authentication. Last, but not least, 802.11h enhances the 802.11a physical and MAC layers to provide network management and control extensions for spectrum and power management.

2.1.2.4 Summary

'Dot eleven', as some people call the IEEE 802.11 standard for short, is a success story: Almost every new notebook ships with 802.11 functionality and several other electronic devices (such as television sets and digital cameras) are following suit. There are multi-mode cards that support different types of 802.11 and products with integrated 802.11 and GSM/UMTS functionality are available on the market.

2.1.3 Cellular Networks: GSM and UMTS

Cellular networks offer wireless connectivity over a wide area, for example within a country or even beyond national borders. The term 'cellular' refers to a partitioning of the area into 'cells', each of which is served by a base station. The fundamental idea behind the cellular concept is that radio channels can be spatially re-used; thus, a given radio channel can be used in different cells at the same time as long as these cells are far enough apart so that simultaneous transmissions do not interfere. Cellular networks are typically operated in a licensed frequency band. The network operator acquires a range of frequencies, (for example, from a governmental institution) and then maintains and operates the network. Users subscribe to the network and can use their mobile phones for voice telephony and data applications.

2.1.3.1 Global System for Mobile Communication (GSM)

The first generation of cellular systems was based on analog radio transmission. A variety of systems were deployed in different countries in the 1980s. The development and standardization of second generation cellular systems – based on digital radio transmission – followed in the late 1980s, with GSM being the most prominent example.

The development of GSM started in 1987, when European network operators and the national administrations signed a common 'memorandum of understanding', which confirmed their commitment to introducing a common digital cellular telecommunication system. The first set of technical GSM specifications was published in 1990 by the European Telecommunications Standards Institute (ETSI); the first networks started operation one year later. In the following years, GSM was deployed in many countries also outside of Europe and became a huge success story.

The main service of GSM is mobile voice telephony with a quality comparable to the fixed network. This is complemented by a number of supplementary services, for example call forwarding and voice boxes. In addition, low-rate data services are offered, in particular the Short Message Service (SMS), which enables users to exchange short text messages in a store-and-forward fashion. From a transmission perspective, GSM uses a combination of Frequency Division Multiple Access (FMDA) and Time Division Multiple Access (TDMA) in the frequency bands around 900 MHz and around 1800 MHz. In the 900 MHz band, frequencies from 890 to 915 MHz are used for the uplink transmission from the mobile device to the base station, while the frequencies from 935 to 960 MHz are used for the downlink from the base station to the mobile station. GSM is thus a system that employs Frequency Division Duplex (FDD) to separate uplink and downlink transmissions. Each band is divided into 124 frequency channels of 200 kHz each. One frequency channel in then again divided into eight time slots. Each mobile device is allocated a given time slot, which leads to a maximum data rate of 9.6 kbit/s in the original GSM.

From a networking point of view, seamless connectivity and mobility are achieved by automatic handover functionality from one base station to another and international roaming agreements between network operators in different countries. For this purpose, the network contains dedicated databases that track the current cell in which a device is located. Security mechanisms include ciphering, protection of the user identity and authentication. An important feature is that a user's authentication data (allowing him or her to access the network) is not stored in the mobile device itself but on a small chip card, called Subscriber Identity Module (SIM), which can be inserted into the mobile device. By exchanging the SIM, people can use different devices while still being reachable on the same single phone number. All in all, the GSM standard defines a complete system and protocol architecture, including solutions for physical transmission of speech and data, medium access, mobility management, radio resource management, services, security, interoperability, network management and other functions.

Since its introduction to the market, GSM has been developed further in a continuous manner. Among other things, voice codecs have been improved to achieve better speech quality and group call and push-to-talk speech services have been implemented. Development also continued with improved data transmission techniques, accounting for higher data rates and packet-based data services:

- The High-Speed Circuit-Switched Data (HSCSD) service achieves higher data rates by allowing a mobile device to transmit in parallel on several time slots of a frequency channel (multi-slot operation). Data rates in the order of some 10 kbit/s can be achieved.
- The General Packet Radio Service (GPRS) introduces a packet-switched transmission at the radio interface. In principle, certain time slots are now allocated in a dynamic manner to different mobile devices rather than being reserved for a particular device for the entire session or call duration. GPRS improves and simplifies wireless access to the Internet. Users of GPRS benefit from shorter access times, higher data rates, volume-based billing and an 'always on' wireless connectivity.
- The Enhanced Data Rates for GSM Evolution (EDGE) replaces the original GSM modulation by a higher-order modulation. It thus enables higher data rates (more than 100 kbit/s) and a better spectral efficiency.

2.1.3.2 Universal Mobile Telecommunication System (UMTS)

Standardization efforts for the third generation (3G) of cellular networks started in the early 1990s, when the decision was made to allocate frequency bands around 2 GHz to future cellular communication. The goal was to develop a family of systems, called International Mobile Telecommunications 2000 (IMT-2000), offering higher data rates than second-generation networks, Internet and multimedia services, high speech quality and efficient spectrum utilization.

Different companies and standardization organizations from Europe, Japan, China, USA and Korea joined their forces and founded the Third Generation Partnership Project (3GPP), which subsequently developed and standardized the Universal Telecommunication System (UMTS). It builds on the existing 2G network infrastructure but introduces a new radio transmission technology, applying Code Division Multiple Access (CDMA). In fact, two different modes are specified:

- a Frequency Division Duplex (FDD) mode with Wideband-CDMA (W-CDMA)
- a Time Division Duplex (TDD) mode applying a combination of CDMA and TDMA

The first worldwide 3G network was launched in Japan by NTT DoCoMo in October 2001 under the brand name FOMA (Freedom of Mobile Multimedia Access). Operators in Europe followed in 2002, while commercial UMTS penetration took place in 2005.

The UMTS air interface enables a mobile device to transmit at a maximum theoretical data rate of 2 Mbit/s; in practice, data rates of some 100 kbit/s are achieved. As a result of the new air interface, UMTS itself is incompatible with GSM. UMTS phones are usually UMTS/GSM dual-mode devices. They include automatic handover from UMTS to GSM (and vice versa) if the coverage of the respective network degrades.

The higher data rates of UMTS give rise to data applications that were not possible or not practical with GSM. In particular, UMTS users can browse the Web and exchange multimedia-enriched messages using the Multimedia Messaging Service (MMS). Other services offered include flat-rate streaming of television channels, video telephony and download of games and music. In fact, the mobile phone has changed into a multipurpose multimedia device that integrates the functionality of a phone, music player, camera, radio, video game, small television set, alarm clock and other devices.

The development toward higher data rates continues with High Speed Packet Data Access (HSDPA). This is a new packet-based downlink technology in W-CDMA systems, which has the potential of offering data rates of the order of 10 Mbit/s. Among other things, HSDPA exploits adaptive modulation and coding techniques as well as multiple antennas on the mobile devices. Deployment of HSDPA started in late 2005 and it is now available from a large number of operators.

2.1.3.3 Service Platforms

An important enhancement of cellular networks has been the introduction of service platforms. Instead of standardizing specific services, features to create services and mechanisms that enable their introduction have been defined. These platforms allow network providers to introduce new services in a fast manner. It thus enables them to differentiate themselves by providing operator-specific, value-added services in addition to the standard services. The GSM/UMTS family defines the following service platforms, which are described in more detail in Chapter 3:

- CAMEL (Customized Applications for Mobile Network Enhanced Logic) is an overlay that introduces the functionality of the Intelligent Network (IN) into GSM. It gives operators the opportunity to offer advanced supplementary services, such as toll-free calls, prepaid calling and private number plans.
- The SIM Application Toolkit (SAT) enables the operator to run specific applications on the SIM card. With SAT, the SIM card is able to display new operator-specific menu items and logos and to play sounds. For example, users can download new ring tones to their phone.
- The Wireless Application Protocol (WAP) introduces a Web-like information platform for mobile users. It defines a system architecture, protocol family and application environment for transmission and display of Web-like pages for mobile devices. Using a WAP-enabled mobile phone, subscribers can download information pages, such as news, weather forecasts, stock reports and local city information. Also mobile e-commerce services (such as ticket reservation and mobile banking) are offered. Another service platform for Web-like information services is i-mode. This was developed by NTT DoCoMo and was originally deployed in the Japanese cellular network, but was then exported worldwide and introduced to GSM networks as well. In contrast to WAP, i-mode was very successful from its beginning. The main success factor was an innovative business model that introduced the role of content providers.
- The IP Multimedia Subsystem (IMS) was introduced in UMTS to control IP-based multimedia services, such as voice over IP. IMS combines voice and data in a single packet-switched network.

2.2 Wireless Networking at a Glance

This section introduces wireless networks, based on cellular networks. The goal is to provide an understanding of a mobile network, as seen from a user or application developer. We explain the main architecture and functions, in particular regarding mobility management, paging and handover, but not complete protocol suites. In this vein, we do not elaborate on the specifics of the wireless transmission technologies and omit many important control functions related to wireless. Examples are power control, radio resource management, antenna technologies and many others.

We show that future wireless architectures can also accommodate different wireless technologies and thus incorporate WLAN hotspots. IP technology provides the basis for gluing these heterogeneous technologies together.

A further extension is to reach out to ubiquitous devices. While, currently, mobile phones, PDAs and portable PCs are the devices connected to wireless networks, in the future many more embedded devices will be equipped with wireless communication systems. Examples are household devices, such as fire detectors, entertainment devices and sensor devices in the public infrastructure, for example temperature sensors in trains. Another important class of devices is passive RFID tags, which can transfer some identity information when triggered wirelessly. These will enable a device to access information in its environment.

In this way, wireless networks will provide the basis of an increasingly digital world. The famous Metcalfe's Law states that the value of a communications network is proportional to the square of the size of the network. The value of a broadcast network is estimated to be proportional to the number of users. This rule is called Sarnoff's Law. It is still to be determined how the value of a network grows with the ubiquitous environment, as many devices are just passive. Furthermore, the value of many pieces of information depends on their accessibility and timeliness, which may be limited for ubiquitous devices. If the growth rate is greater than linear, but below quadratic, the sheer number of devices will lead to an enormous growth in the value of the network.

2.2.1 Cellular Network Architecture

As mentioned in Section 2.1, wireless networks are often based on the cellular principle. This section discusses the system architecture of such networks, as incorporated in second- and third-generation cellular networks like GSM and UMTS. A simplified view of a cellular network is shown in Figure 2-4. It is structured into an access network and a core network. From a networking point of view, both parts have the goal of providing a seamless service. This means that they create a non-interrupted service over an inherently fragile wireless link, further aggravated by the need for handover between base stations in case of mobility.

The **access network** has the following responsibilities:

- wireless attachment of mobile devices to the network
- local mobility management including paging to locate devices whose location is unknown
- handover of ongoing calls or sessions between base stations
- radio resource control, i.e. management of physical-layer resources of some part of the access network
- link-layer security, including procedures for user identification and access control

As shown in Figure 2-4, the tasks of the access network are implemented in the base stations and radio network controllers (RNC).

The **core network** has the following responsibilities:

- global mobility between networks or network segments, including reachability of nodes by an address that is location independent (for example, a phone number)

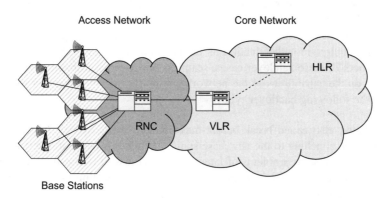

Figure 2-4.   Second-generation cellular network

- control of user credentials and admission to network
- roaming support for services between networks
- routing of data packets and call signaling, i.e. traffic on user and control plane, including issues like name resolution
- providing interfaces for service control to other parties, e.g., for call setup initiated by some external entity

For a good overview of terminology in this area we refer to RFC 3753.[1]

2.2.2 Mobility Management and Handover

Mobility management refers to the capabilities of a network to cope with the mobility of users. We can distinguish between seamless mobility and nomadic mobility. The former provides a service continuously while a user is mobile. The latter refers to mobile nodes that can attach to a network at different locations, but do not expect continuous reachability and seamless services.

The main technical problems of mobility management are thus the following:

- reachability of a mobile device at different points of attachment to the network
- seamless handover of ongoing services between base stations in case of device mobility

The reachability is resolved using a fixed address that is not affected by mobility and a lookup service to discover the current location or temporary address for a mobile device. In GSM systems, the home location register (HLR) and the visiting local registers (VLR) contain the current location of a mobile device, such that calls can be forwarded to this location (see Figure 2-4).

Seamless handover means that a session should not be noticeably interrupted if a user moves from the area of one base station to the area of another base station. There has been considerable research on reducing the service interruption of a handover to a new base station. Most critical are interactive services such as voice calls, which are highly susceptible to loss of data.

We can classify handover types by the technologies used and the administrative boundaries. A handover is called an 'inter-technology handover,' if the old and new networks employ different wireless or networking technologies. Another type is 'inter-domain handover' between different network domains, for example belonging to different network operators. This typically requires re-authentication as the credentials are not exchanged between the networks. In both of these handover cases, seamless handover is difficult.

Depending on the capabilities of the network and on the specific handover incidence, we can classify the following handover types:

- Hard handover, also called break-before-make handover, means releasing the old base station and then attaching to the new base station. While this is the simplest kind of handover, it can also lead to problems of latency and data loss. It is also the most basic case, because, in wireless communications, losing the old base station will generally happen without advance notice.
- Make-before-break handover requires using simultaneous connections to both target and old networks. In this way, data can be delivered over both paths at the same time and packet

loss can be reduced to virtually zero. The practical problem is that in many mobile systems, devices are limited and cannot connect to two base stations at the same time, for example because they use different frequencies.

- Anticipated handover, also called planned handover, is a 'prepared' hard handover, which means that the connection to the new base station is already prepared before the connection to the old one is broken. For instance, registration at the new base station, which needs some exchange of credentials, can be prepared.

The concept of anticipated handover can avoid problems of finding a suitable base station and allocating resources at the new base station and its underlying network. This is particularly useful for inter-technology handover and possibly also for inter-domain handover. In both cases, anticipated handover requires information about the new networks in advance so that the most suitable network can be selected. Furthermore, some approaches use the old network to get in touch with the new network.

There are different ways to support make-before-break handover in the network. In the simplest case, the mobile device may be able to connect to two networks simultaneously. This may be possible if the device connects to two networks with different technologies and hence has two independent transceivers. This is called network diversity and support is needed to pass an ongoing session efficiently from one network to the other. Another case is macro-diversity, where the physical layer supports such handovers. For instance, in CDMA-based systems, devices can communicate with two base stations at the same time and this kind of handover is also called soft handover.

2.2.3 Paging

A further concept that is intertwined with mobility management is paging. To route incoming calls and data sessions, the network needs to know the location (cell) of the mobile device. This requires the device to send location update messages to the network. Since mobile devices are battery-limited, sending messages to base stations should be as efficient as possible. To reduce signaling overhead and energy consumption in the devices, several cells are combined in a paging area. A device in its idle state does not always send a location update when it changes to a new cell, but only when it changes to a new paging area. When an incoming call is received for a device, paging is performed in all cells of the device's current paging area, i.e. all base stations in these cells query for the mobile device on a separate signalling channel to determine the location of the device.

There are several different concepts used to implement paging. Two example strategies are blanket polling and sequential paging. In the first, a paging request is sent in all cells of a paging area simultaneously. In the second, paging requests are sent to base stations sequentially in decreasing order of the likelihood that the device is located in that cell. All strategies trade off the effort that devices have to make in location updates with the effort that the network has to make to contact nodes for incoming calls.

2.2.4 Roaming

Roaming agreements between operators of different networks enable users to roam between these networks. This enables mobility between networks, although this is often limited to nomadic mobility. In technical terms, roaming requires a foreign operator to admit a user without direct and full access to the user credentials. The reason for this is that operators typically cannot (and do not want to) exchange the shared secrets or secret keys of their users with other operators. Thus, the foreign operator can only delegate admission control to the home operator, or it receives limited information to enable it to perform this check. A typical way of implementing this is to request a challenge and answer pair from the home operator. With this pair, the foreign operator can send the challenge to the mobile device, which then uses its internal secrets to produce a correct reply, which is then checked by the foreign operator. In this way, only one round-trip message exchange with the home network is needed.

In addition to user authentication, operators need to agree on the services for which a user is authorized and also on the charging between the operators. This is done by bi-lateral agreements, called service level specifications (SLS), which are set up in advance.

2.2.5 Quality of Service

Quality of service (QoS) is a widely used term in networking and wireless systems. While the term 'service' should refer to an end-user service, the term 'quality of service' is commonly used for different and more specific technical issues.

Quality of service in a layered system model, as in most communication systems, refers to a specific layer employed by the systems. For the application layer, this refers to the actual quality perceived by the user. This is clearly application service specific and, for example, measures the quality perceived by human users. For applications, it is often quite difficult to translate measured quality into technical parameters. For instance, consecutive data loss is typically more severe than just sporadic loss for typical voice codecs. Furthermore, these measurements depend on many issues such as the language used in voice communication. Other QoS parameters on the application level are easier to measure, for example call setup time or call drop rates.

On the network layer, QoS typically refers to the end-to-end transport capabilities available for a user service. This includes parameters such as data rate, delay and loss. As application requirements may vary dynamically, and also because networks typically have service variations, statistical operators, such as maximal variation, average loss or maximum data rate, are used. To structure the network services, applications are typically classified into several classes depending on their requirements. A typical example is the ITU-T recommendation Y.1541 of QoS classes, as shown in Figure 2-5. On the link and physical layers, quality of service is measured in terms of bit error rate or the capacity of a point-to-point link.

Most current cellular systems are optimized for voice quality of service. This means that delay and error handling on different layers (including voice codecs) are optimized for voice traffic. Considerable efforts are also made to provide suitable quality of service for data services. For these, the main problems are that the required data rate is typically not known in advance and that the requirements for call or session setup are much higher. While the setup time for voice calls can be several seconds, or up to one minute in rare cases, such a high start-up delay is usually unacceptable for data communication. In particular, techniques such as paging are suitable for voice calls but may create long session setup times that are not suitable for data applications. On the other hand, many data applications are less susceptible to loss or delay, once the session is initiated.

	Class 0	Class 1	Class 2	Class 3	Class 4	Class 5
Delay	100 ms	400 ms	100 ms	400 ms	1 s	Unspecified
Delay variability	50 ms	50 ms	Unspecified	Unspecified	Unspecified	Unspecified
Packet loss	10^{-3}	10^{-3}	10^{-3}	10^{-3}	10^{-3}	Unspecified
Error ratio	10^{-4}	10^{-4}	10^{-4}	10^{-4}	10^{-4}	Unspecified
Example services	Voice, video	Voice, video	Signalling	Interactive data, signaling	Streaming video, bulk data	Best effort data

Figure 2-5. QoS classes and thresholds as specified in Y.1541

2.2.6 Location Services

While the inherent goal of cellular networks is to provide communication services, they can also provide location services. In other words, the location of a user can be determined by the network, typically by triangulation of the signal from a device as received by several base stations. The accuracy of this information varies depending on the network coverage. The operator can use the location information itself or forward it to third-party service providers.

Another location system is the Assisted Global Positioning System (A-GPS), which is a variant of the Global Positioning System (GPS) used in cell phones. Here, a device receives information about GPS satellites (orbit, frequencies, and functionality) from the cellular network. As a result, the device's GPS receiver can detect and analyse even weak satellite signals. Compared to traditional GPS systems, A-GPS is faster and saves battery power.

For both technologies the network operator or a third party can exploit the location information to offer location-specific services (for example navigation services, location-based emergenccalls or recommending nearby restaurants).

2.2.7 Broadcasting and Multicasting Services

An attractive service in communication systems is one-to-many communications, for example for content broadcasting or multicasting of a session. The motivation for the network is that resources, especially on the wireless link, can be saved by simultaneously sending data to several destinations. Furthermore, it is also convenient for the sender, as it can reduce the number of sessions it has to transmit. For instance, for a video server, this can increase its scalability significantly. On the other hand, one-to-many distribution assumes that all users want to receive the same content at the same time.

Multicasting can be separated into two functionalities: multicast data distribution and group management. Two major standards are Internet multicast protocols[2] and the 3GPP multimedia broadcast and multicast services – MBMS.[3] The latter extend parts of the former by optimizations for wireless networks and by integration into the UMTS system architecture.

There are several issues for multicasting in mobile networks. First, mobility of users must be accommodated for the distribution groups. Second, the media distribution to a number of users must be supported efficiently by appropriate wireless technologies. This is however non-trivial; for example, power control is specific for each terminal, which cannot be done in the same way for a group of nodes.

2.3 Next-Generation IP-based Mobile Networks

The trend in fixed and wireless networks is towards a dominance of Internet technology with all its flavours. In this section, we discuss the motivation and some technologies for next-generation mobile networks that will be based on the Internet protocol (IP).

As we have seen in Section 2.1, there are a variety of wireless technologies that complement each other regarding coverage area and data rate. A major issue for next-generation networking is providing seamless radio access in this heterogeneous scenario. Here, IP serves as the 'lingua franca' for enabling the exchange of data across various technologies, such as WLANs, cellular networks and upcoming 4G wireless technologies. To support heterogeneous access, mobile networks are moving towards all-IP networks, aiming for both extensible and cost-effective operation. A schematic view of next-generation IP-based mobile networks is shown in Figure 2-6.

The Internet protocol suite has largely been successful because it follows some basic principles. The Internet offers basic connectivity as the minimal service without any service guarantees, and this turned out to be simple and effective for the World Wide Web and for data applications. The Internet architecture is based on the end-to-end principle, which gives the end systems as much control as possible. This reduces the state in the network, which leads to reduced cost of the network and improves the flexibility for installing new functions

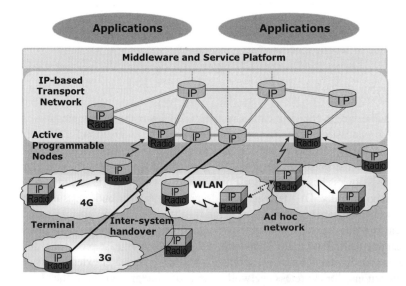

Figure 2-6.   IP-based mobile networks

and applications on the terminals. As the data rate in fixed networks has become more afford-able at a speed that outpaces the usage, the simple networking technologies of the Internet have succeeded by over-provisioning. For a discussion of the Internet architecture we refer to RFC 3724.[4] The design of the Internet and its impact on social, economic and political aspects is discussed by Clark *et al.*[5]

There are several reasons for mobile operators to migrate to a single IP-based networking solution. First, IP offers packet-based transport, which is flexible for different applications. Second, IP is a widely used networking technology that is independent of the underlying wireless technology.

On the other hand, any future mobile Internet architecture has to address the deficiencies of the current Internet, where mobile networks and mobility aspects are treated with com-paratively low priority. As mobile networks have to handle many additional functions con-cerned with user mobility and scarce wireless resources, the mobile Internet will have to accommodate several new functions.

2.3.1 Heterogeneous Access and Mobility

The multiple networks provided by different operators, together with the associated technolo-gies, lead to increased affordability of wireless communication, because the user has full freedom to select the technology and service offering. In addition, investment costs for new networks are reduced. On the other hand, networks will have to integrate the capabilities of different technologies to provide an end-to-end, seamless and secure solution for the user.

There are many implications of heterogeneous wireless systems for the overall network, including services and media delivery. The main challenge is to support seamless, optimized services over different wireless access technologies. To offer user-oriented services in het-erogeneous environments, the network has to act intelligently on behalf of the user to optimize the network services. For instance, we envision the use of context information about the user, the network and its services in order to provide the best network service. In this way, networks will offer seamless and user-oriented services, and the user will not need to be aware of dif-ferent radio access networks.

As an example, a handover to a more suitable access point offering more capabilities may be needed to enable additional services. For instance, when passing a wireless hot spot, one could perform a handover to this access point for a short period of time so as to facilitate some demanding service, such as download of bulk data or video conferencing. However, in many cases the availability of resources at the potential access point will not be known before the handover is performed.

2.3.1.1 IP-Based Mobility Management

Originally, the Internet was designed for a permanent attachment of hosts, where users with fixed IP addresses are bound to a specific network location. In this sense, an IP address serves both to identify a node in the network and to determine the route to this node. As mobility requires separating locators, used for routing data, and identifiers, there are several approaches to extending the Internet protocols.

The most prominent example is Mobile IP,[6,7] which works with two IP addresses. One address, the home address, is used to identify the device while another address, called care-

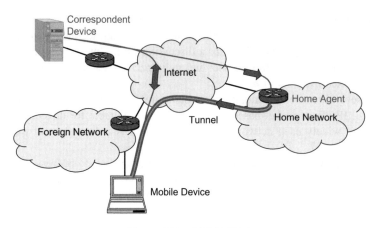

Figure 2-7.   Mobile IPv6

of-address, describes the current location of a device. Packets destined for a mobile device are first sent to its home address, where a home agent provides forwarding functionality to the current care-of-address.

The basic mechanisms of Mobile IP are illustrated in Figure 2-7. The main new network entity is the home agent, which resides in the home network of a mobile device. Mobile devices have to register their care-of-address with the home agent. The home agent can intercept incoming traffic for this device and forward it to the device. To implement this forwarding, tunnelling of data packets is needed.

The advantages of Mobile IP are the migration from existing networks, as only one new network entity is needed. There are no changes to existing TCP/IP applications or to the protocol stack in other devices. Also, as a result of the tunneling, the routing is not affected. As IP is independent of the wireless technologies, Mobile IP provides a solution for heterogeneous networks.

On the other hand, Mobile IP has drawbacks, which gave rise to extensions and alternative solutions. First of all, the data path is not optimal, although there are several optimizations for more direct routes, for example route optimization in IPv6. This requires that the corresponding device be notified of the care-of-address and needs to keep track of the other device's mobility. In turn, this means that the correspondent device will learn about the location or movement of the mobile device; this gives rise to location privacy issues.

Another disadvantage of basic Mobile IP is that it does not support efficient handovers, as all location update traffic has to go to the home agent. In this sense, it more or less corresponds to the functions of the mobile core networks discussed above, as the focus is more on reachability. Because of the insufficient support for frequent local handovers in cellular environments, IP mobility solutions have been separated into two parts:

- Macro-mobility protocols, such as Mobile IP, implement the reachability of users on a large scale, between different wireless networks connected over the Internet.
- Micro-mobility protocols are used to manage mobility within a visited domain and are optimized for frequent handovers. A number of different micro-mobility protocols are currently being developed.[8] These approaches have the advantage that local movements are

not visible outside the domain, which in turn reduces handover latency and signaling load. On the other hand, these approaches create a more mobile-specific state and require additional security associations between the nodes and the network.

Another issue of an open network like the Internet is that mobility gives rise to a number of security threats. For instance, a device may hijack a session during a handover or during route optimization. In both cases, a malicious device may listen to or intercept traffic and then pretend to be another mobile device and divert traffic to itself.[7] The main challenge has been to address these issues without relying on a security infrastructure within cellular systems. These drawbacks of Mobile IP are addressed by some more recent approaches.

An alternative approach to IP mobility management is the IP² architecture,[9] where the network takes full control of mobility management and the nodes are not concerned with mobility. In contrast to Mobile IP, the IP² architecture aims at defining a new mobility management layer consisting of routing management and a newly defined location management. This should improve several issues of mobile IP, including route optimization and location privacy. On the other hand, this approach deviates from the traditional Internet architecture, which is based on an end-device-centric view based on the end-to-end principle.

Another recent, more basic approach is to introduce a new naming layer between the IP layer and the applications. This is pursued by the IETF in the Host Identity Protocol (HIP).[10] Here, cryptographic identities are used for this new naming layer. Two devices that want to communicate first establish a security association between them. On the one hand, this can be used to avoid denial-of-service attacks and is also useful for mobility control. On the other hand, the concept deviates from the original Internet, which provides data transport without any session setup.

2.3.2 IP Quality of Service

Quality of Service (QoS) for the Internet has been a research issue for about 15 years and several solutions have been developed and standardized. In the following, we introduce some major QoS approaches, where we distinguish between the data plane and the control plane. On the data plane, sometimes called the user plane, QoS has to ensure that the actual data packets for one service class or data flow receive the agreed service, either guaranteed or statistically. The control plane has the responsibility for configuring the data path mechanisms and typically provides some signalling to accept user requests for QoS services and to provision the network accordingly unless a request is rejected.

In the data plane, the **Integrated Services (IntServ)** QoS mechanism is based on end-to-end per-flow resource reservation. A flow is defined as a data stream generated by one application, roughly specified in terms of source/destination addresses and port numbers, which indicate the application. Different levels of guarantee are specified for guaranteed services and controlled load, where the latter gives the softer statistical guarantees. End-to-end reservation has to be established at all routers on a given data path. This is done by the control plane, initiated by a signalling protocol such as RSVP.[11]

The second prominent QoS mechanism defined in the IETF is **Differentiated Services (DiffServ)**. Instead of the per-flow state at each router in a network, QoS preferences or guarantees are assigned to traffic aggregates or classes, which are composed at the network edges. This is implemented by marking packets in a special field in the IP header, the DS field. Such a mark is called the DiffServ Code Point (DSCP), which provides information about the QoS requested for a packet. There are a few bits available for DiffServ Code Points and several so-called per-hop behaviours (PHB) are specified for these. This means that the treatment of the packets in a class is only specified for a single hop, not end-to-end. Hence, end-to-end services must be constructed on top of DiffServ – the DiffServ model does not include an end-to-end service. The management of aggregates instead of flows leads to a better scalability for DiffServ, at the price of no strict guarantees for each flow.

The two primary DiffServ per-hop behaviours defined by the IETF are the Expedited Forwarding (EF) PHB[12] and the Assured Forwarding (AF) PHB. The Expedited Forwarding (EF) PHB implements a virtual leased-line service where the bandwidth cannot exceed some guaranteed peak rate. However, the actual implementation of the EF service requires additional ingress traffic control at the edges of the network. Also the state in the network along the path of the leased-line service must be installed. These mechanisms are not standardized by the IETF and may be implemented with Bandwidth Brokers or some other signaling, as discussed below.

The Assured Forwarding (AF) PHB[13] does not provide a bandwidth guarantee, but packets are given different priorities at each hop. In congestion situations the user of the Assured Forwarding service should encounter less bandwidth decrease than Best-Effort users.

Four Assured Forwarding service classes are defined, which are typically implemented by different queues in the routers. Each of these classes has in addition three levels of dropping precedence: low, medium and high. This dropping is orthogonal to the classes and can differentiate packets in one queue. The advantage is that packets in the same class, but potentially with different drop priorities, are not reordered. This is relevant for many services such as multimedia content, where often some data is less significant than other, but the ordering is important.

Another QoS mechanism is **Multi-Protocol Label Switching (MPLS)**. This is a layer 2.5 protocol where a flow is identified by a so-called flow label, which is inserted between the layer 2 and layer 3 headers. MPLS provides routing for a flow and is used for both traffic engineering and QoS. For traffic engineering, the flows can be forwarded independently of the IP routing, which gives the network designer some control over the actual data flows in the network. For QoS, there are extensions that allow to specify QoS parameters for each MPLS flow. MPLS attempts to overcome the scalability issues of the per-flow state by the flow labels, which permit simple hardware support. This is similar to virtual circuits in ATM networks from where the concept originates.

Regarding control planes for quality of service, we can distinguish two main classes, hop-by-hop and off-path signalling.

The **hop-by-hop reservation** approaches are based on end-to-end signaling along the data path of a data flow, also called on-path signalling. The hop-by-hop reservation architectures are characterized as follows:

- QoS resources are managed locally by each router.
- Signalling is triggered by the terminals and follows the data path.
- End-to-end reservations are set up hop-by-hop by stateful mechanisms in each router.

The widely discussed weakness of the hop-by-hop model is its scalability, due to the per-flow state and processing in each router. There are three major representatives of this architecture, namely RSVP, MPLS signalling, and NSIS. RSVP was designed as the control protocol for IntServ and relies on the IntServ flow concept. It is a layer 3 protocol that establishes the state in each router of the data path and is flexible with respect to multicast extensions and data path re-routings. Recently, work on QoS signaling has been started by the IETF in the 'nsis' working group.[14] The goal of this working group is the definition of more generic signalling protocols that can support different QoS techniques as well as other signalled functions.

The **off-path QoS control architecture** is based on dedicated entities in the networks, which we call the Resource Manager (RM). This is also known as a bandwidth broker[15,16] and is characterized as follows:

- A single Resource Manager (RM) handles the resources for each domain.
- The RM maintains an up-to-date image of resources and reservations in its domain.
- The resources can only be reserved via the RM, which performs admission control.

The advantage of this concept is that fewer network entities must be invoked when a QoS request is received. In particular, for non-guaranteed service, there is no need to establish state in each router. Hence the RM may just monitor the overall network congestion and base its admission control on this. Alternatively, the RM may implement guaranteed services based on the EF DiffServ model. In this approach, the traffic is controlled at specific ingress points, called edge devices, and assigned a PHB. For each link in the network, some resources must be reserved for EF and the RM must check whether these resources are used by the flows that it admits.

The potential drawback of the off-path signalling solutions is that signalling is not coupled with the data path. This means that failures on either of the signalling or data paths are unrelated, and hence make the error handling more difficult.

While several approaches have been considered, so far no support for a QoS scheme has been widely deployed in the fixed Internet for end-to-end services. As bandwidth in fixed networks is usually cheap, best effort with suitable provisioning is mainly used currently. This is generally sufficient for web surfing and possibly for low-bandwidth applications such as voice calls. There has been considerable discussion why IP QoS has not been adopted[17,18] – major issues include business models and timing of approaches regarding market demand. As we progress towards more demanding applications and towards mobile networks, we expect that the demand for QoS will increase and also that revenue models can be implemented.

2.3.2.1 Mobile Internet Quality of Service

The main requirements for QoS signalling in mobile IP-based networks are as follows:

- interworking with different mobility concepts for seamless handover, including inter-domain handovers
- independence of a particular QoS technique for providing QoS on the data path
- independence of specific radio access technologies

Using the Internet protocol implicitly fulfils several of these requirements. A particular issue is the integration of mobility support and QoS. An overview of some approaches is given by Manner *et al.*,[19] with a classification of the interaction between resource and mobility management.

The QoS signalling approaches require considerable extensions for them to support different handover models. For instance, with on-path signalling, all routers in the old and new paths have to be coordinated in the case of a handover as they contain state information for a data flow. In the case of anticipated handover, the new data path from the anticipated access point to the corresponding node has to be determined and resources have to be reserved in advance. In Figure 2-8, we illustrate the QoS problem of anticipated handover.[20] Figure 2-9 shows the case of off-path signalling using a RM for each domain of the network. In this

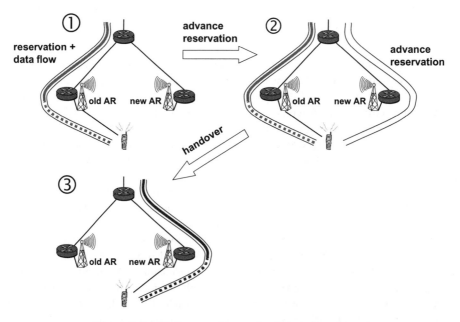

Figure 2-8. QoS reservations and anticipated handover

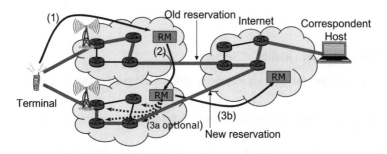

Figure 2-9. Domain-based QoS control and anticipated handover

example, signalling over the old base station (1) can be used to inform the new network (2) about preparation of resources for an upcoming handover that takes place in the next step (3). After a successful handover, the old reservation may be released. In this way, we support pre-reservation of resources before attaching to the new network, either triggered by the mobile node or by some network intelligence (for example, movement prediction).

2.4 Ubiquitous Communication and Ad Hoc Networking

Wireless connectivity of phones or notebooks has become very important over the last few years. The great success of GSM and the wide-spread deployment of WLANs are two prime examples of this development. Now, more and more portable devices are becoming networked, following the trend toward ubiquitous computing. Ubiquitous computing means that computers are embedded into everyday items – such as electronic books, identity cards, refrigerators, cars and washing machines – enhancing them with advanced functionalities. The vision of ubiquitous computing is that the computer actually disappears into the physical environment around us and that all items are networked with wireless technologies. This vision was recognized very early by Weiser,[21] who argued that computing devices will be embedded universally and hence that the traditional computer is disappearing.

Ubiquitous computing creates not only new applications but also demands for new technologies and protocols for wireless communications. Here, the primary concern is not necessarily a very high data rate, but energy efficiency, miniaturization, distributed network organization, scalability and low device cost are typically of major importance. Wireless networks of ubiquitous computers are also expected to be more dynamic and heterogeneous than today's wireless networks. The cellular principle of GSM and UMTS, where mobile devices always communicate with a base station and the network infrastructure is centrally managed, is not well suited to ubiquitous computing. In contrast, we need wireless technologies that enable the devices to communicate directly with each other in a peer-to-peer fashion. It should be possible to form networks among devices in a spontaneous manner ('on the fly') without any cost or effort of building up and maintaining a network infrastructure. This paradigm is followed by an emerging technology for mobile communications, called 'ad hoc networking'. In an ad hoc network, mobile devices establish a self-organizing wireless network without the need for base stations or any other pre-existing network infrastructure. An outstanding feature is wireless multihop communication: if two devices cannot establish a direct wireless link (because they are too far away from each other), devices in between act as relays to forward the data from the source to the destination. Wireless multihop networking in which most devices are stationary is commonly referred to as 'mesh networking'.

The following sections describe some application scenarios for ubiquitous, ad hoc and mesh networking. We cover spontaneous (ad hoc) networks among mobile computers, wireless sensor networks, wearable computing and networks among vehicles.

2.4.1 Mobile Ad Hoc Computing

Perhaps the most straightforward application of ad hoc networking is establishing spontaneous wireless connectivity between several mobile computers, for example to work on a

Figure 2-10. Ad hoc mobile computing scenario

common document at a conference or to access shared equipment such as projectors and printers (see Figure 2-10). Introducing multihop capability to such wireless connections makes it possible to scale the network to more computers and a larger area. For example, a campus-wide ad hoc network can be established in which messages are transmitted in a hop-by-hop manner from computer to computer. Other applications are found where infrastructure is not available, too costly or inefficient to use, or where it has been damaged. For example, imagine that a natural disaster happened in a certain area that also destroyed the technical infrastructure. Humanitarian organizations and other rescue teams would require robust net-working support to coordinate medical aid and clean-up efforts. Ad hoc networking can make such support possible.

The development and standardization activities to make these scenarios and others become reality are ongoing. The Internet Engineering Task Force (IETF) has developed a set of IP-based routing protocols for mobile ad hoc networks. Among them are the protocols Ad hoc On-Demand Distance Vector (AODV), Dynamic Source Routing (DSR), Optimized Link State Routing (OLSR) and Topology Dissemination Based on Reverse-Path Forwarding (TBRPF). While the underlying routing mechanisms have some similarities with conventional IP routing, the ad hoc protocols react immediately to failures or mobility of devices. Com-pared to mobility management in cellular networks, they usually operate in a completely decentralized manner, that is, there is no central location register or other database. Since the forwarding tables for the routing decisions are derived from information about the topology of the network, these protocols are commonly referred to as topology-based routing. An interesting routing alternative, which is not yet undergoing the process of standardization, is geographic routing. This differs from topology-based routing in that the routing decisions are based on local information about the positions of neighbouring nodes and the position of the final destination. Packets are then forwarded to a neighbour, who provides forward progress toward the destination (see Figure 2-11). This approach requires that nodes are aware of their position. It has the advantage that, in contrast to topology-based routing, it is not necessary

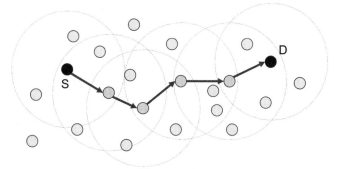

Figure 2-11. Geographic routing from source node S to destination D

to maintain route information at the network nodes, which is particularly beneficial in highly mobile networks. Such protocols are very suitable for networks where location information is readily available (for example for car-to-car communication as discussed in Section 2.4.5). Geographic routing can also be used for the mesh networks discussed in the next section. However, geographic routing only works well if the physical distance between nodes is a good indicator of the connectivity among them. In the more complex radio propagation environments that may be found in mesh deployments, routing based on a virtual coordinate system that reflects connectivity rather than geographic position is likely to perform better.[22]

Much of the work discussed above concentrates on 802.11 as wireless medium, but Bluetooth is also an enabling technology for small-scale ad hoc mobile computing.

2.4.2 Multihop Radio Access Networks and Mesh Networks

Wireless mesh networks are usually considered to be more structured and stable than pure ad hoc networks. These are composed of a mix of fixed and mobile wireless mesh clients and routers. The mesh routers form a backbone to forward packets on behalf of other nodes.[23] They may also provide gateway functionality to other types of network.

Mesh networks are a promising solution for scalable and ubiquitous high-speed communication. Harnessing the benefits, however, is not straightforward. As a result of interference, the capacity of large-scale ad hoc networks is limited, leading to a rapid degradation of throughput for communication over long routes.[24] Among the technologies for improving mesh network performance are the concurrent use of multiple radio interfaces, beam-forming antennas, multiple-input multiple-output (MIMO) techniques and opportunistic channel selection.[25] All of these techniques aim at increasing communication diversity and in turn increase capacity.

Within the IEEE, various working groups are concerned with mesh networking. The goal of 802.11s is to optimize and extend 802.11 to support ad hoc and mesh networking. It mainly enhances the MAC layer with functions for mesh topology discovery and path selection, and includes new functionality to make the medium access efficient for multihop networking. Security functions and mesh measurement are also part of 802.11s. Similar efforts are under way for MANs within the IEEE 802.16a working group and for PANs within IEEE 802.15.5. Further ongoing activities within the IETF address the autoconfiguration of ad hoc or mesh networks.

In parallel with standardization efforts, mesh networking solutions have been developed based on non-proprietary technologies, for example the Roofnet project of the Massachusetts Institute of Technology, and as proprietary solutions from a number of companies (for example LocustWorld, Tropos, BelAir and Firetide). According to the *MuniWireless Magazine* (http://www.muniwireless.com), more than 50 city-wide or county-wide mesh networks have been deployed in the USA to date, and many other countries are following suit.

If the main purpose of a mesh network is the interconnection of an ad hoc network with one or more access points connected to a fixed network infrastructure (for example the Internet), it is often referred to as a 'multihop radio access network'. In such a network, some devices have a direct link to an access point and relay traffic from other devices, which do not have a direct link to the access point. The network topology is typically arranged as a tree with the access point at the root. In this way, the radio coverage of an access point is extended over a larger area.

2.4.3 Sensor Networks

The wireless connectivity of sensors gives rise to many new applications that go well beyond the scope of 'classical' telecommunications. One of the main areas of sensor networks is environmental monitoring. Sensors are placed in an area that needs to be observed. Each sensor has certain sensing tasks, for example to measure the temperature and humidity, to record sounds or to detect vibrations. The measured data is analysed and shared with other sensors, and the most meaningful data is forwarded to dedicated sink nodes, which can be accessed remotely.

Applications can be found in fields such as:

- biology (for example animal observation, glacier monitoring, ocean water monitoring and grape monitoring)
- civil and industrial engineering (for example intelligent buildings, monitoring of bridges and automation)
- medicine and healthcare (for example patient monitoring)
- emergency response and military (for example disaster warning and vehicle tracking)

For ease of deployment, sensor devices should be inexpensive and small, and have a long lifetime, which makes it important to develop very efficient software and hardware solutions. For this reason, protocols for sensor networks should be carefully designed so as to make the most efficient use of limited resources in terms of energy, computation and storage. These restrictions are likely to remain in the future, since in many cases it is desirable to exploit technological improvements to develop smaller and more energy-efficient devices, rather than making them more powerful.

Technical solutions for wireless sensor networks are very application-specific and span a wide range of different implementations. The standards mentioned in Table 2.1 for wireless personal area networking are applicable to sensor networks as well.

The IEEE standard 802.15.4 is a radio technology for low-rate personal area applications. The main benefits are low power consumption and the low cost of the devices. As in WLANs, devices transmit in the unlicensed ISM band at 2.4 GHz. The physical layer in this band defines 16 channels, each having a 250 kbit/s data rate. In some countries, two additional ISM bands at 868 MHz (Europe) and 915 MHz (USA and Australia) can be used. They provide 10 channels with 40 kbit/s and 1 channel with 20 kbit/s. In all bands, a Direct-Sequence Spread Spectrum (DSSS) scheme is applied. The medium access control is based on a Carrier Sense Multiple Access protocol with Collision Avoidance (CSMA/CA), which can be operated either in slotted or unslotted mode. The topology formation is self-configuring, where the network can either form a single-hop star topology with a coordinator device or a multihop peer-to-peer topology. On top of IEEE 802.15.4, the industrial alliance ZigBee has defined a set of higher-layer protocols. The ZigBee networking layer includes mechanisms to join and leave a network, and discovers and maintains routes between devices. On top of the networking layer, an application layer framework has been defined. Among other things, it performs election of coordinators and discovery of services that the devices provide. Last, but not least, the ZigBee platform defines a set of security services for the MAC, network and application layers. Commercial products based on 802.15.4/ZigBee include, for example, the MICAz sensors available from Crossbow Technology and the EM radio chips from Ember.

Another radio technology for sensor networks is Ultra-Wide-Band (UWB). Here, signals are transmitted as very short impulses, thus occupying a very wide frequency band. Although this band typically overlaps with the bands of other wireless technologies, the wide spreading of the signal enables receivers to detect and decode the signal. The major advantages of UWB are as follows: transmissions can be performed with very low power; the data rate is very high at small distances; and the transmission is very robust against frequency-selective fading.

Alternative radio technologies are proprietary solutions developed by companies and within academia. In some product developments, Bluetooth is used as the transmission technology for sensor networks. One well-known example is 'BTnodes' developed at ETH Zürich. Non-radio transmission technologies include optical and ultrasound communication. The latter is especially relevant for underwater scenarios where radio communication is not applicable.

From a networking point of view, wireless sensor networks have several similarities with mobile ad hoc computing networks, such as decentralized control and the possibility of multihop communication. One of the major differences is that sensor networks follow a data-centric paradigm, i.e. the measured data is manipulated as it travels through the network. Since data usually flows from the sensor nodes to one or sometimes several data-collection nodes, sensor routing protocols usually create a tree topology with the data-collection node at the root. Because of the comparatively limited capacity of sensor networks and the energy constraints of the sensor nodes, aggregating and compressing the sensed data is essential. Trading off local computation for a reduction in the number of required transmissions allows these scarce resources to be conserved.

2.4.4 Wearable Computing

The main idea behind wearable computing is to wear small and light computing devices and equipment on one's body, in a similar manner to eyeglasses and clothing. The worn devices interact with the user, taking into account the current context of the user and their environ-

Figure 2-12.   Wearable computer integrated in a belt and attached to head-mounted display

ment. Typically, a wearable computer has a main unit that is integrated into clothes or a belt. It is connected to user-friendly input devices (such as gloves, a camera or a microphone) and output devices (such as a head-mounted display or headphones). Tools for context sensing and communication complement the system. An example is shown in Figure 2-12.

An important feature of wearable computing is the communication between all the parts of a user's wearable computer or between different wearable computers. In principle, we can employ existing technologies for personal and local area networking for this purpose. For certain real-time multimedia applications, however, current radio technologies do not provide sufficiently high data rates. For this reason, alternative non-radio technologies are currently under development. One idea is to use the surface of the human body as a medium for transmitting signals. To be more precise, the natural electric field that surrounds the body can be used as a transmission link with very high data rates. Technologies based on such a transmission principle are still under development. A well-known example is the RedTacton system developed by NTT.

The scope of applications for wearable computing is very wide. The advantages compared to conventional notebooks and pocket computers become most obvious in certain fields of work. A typical example can be found in the area of emergency response, for example a crew of firefighters equipped with belt-mounted wearable computers, head-mounted displays and cameras. The worn computing and processing power enables them to coordinate, to send alarms and to detect changes in the environment. Other scenarios where wearable computing at work is useful are industrial production, maintenance, medicine and healthcare. In the consumer field, wearable computing and communication will enable new forms of entertainment, communication, navigation, and capturing and archiving personal experiences. 'Light versions' of wearable computers are already wide-spread today, namely camera phones and MP3 players.

2.4.5 Vehicular Networking

The integration of wireless technologies into cars gives opportunities for new safety and communication features inside vehicles. For example, wireless ad hoc networking will enable efficient accident warnings, where cars involved in an accident can send warning messages back over a certain number of other vehicles, thus avoiding motorway pileups. Other safety services include sensor-based aquaplaning or traffic-jam warnings. Finally, we can also envision person-to-person applications using ad hoc communication between vehicles (for example simple text messaging, game communities, and hop-by-hop telephony).

There are several ongoing development and standardization activities in this domain. The baseline for all wireless communication is the allocation of frequencies. The safety-related vehicular services mentioned above should not suffer from interference from other wireless technologies and hence require dedicated frequencies. Such frequencies have been allocated in the 5 GHz band in North America and Japan, but have still not been allocated in other parts of the world. To define a physical and MAC layer for such dedicated short- and medium-range communication, the IEEE working group 802.11p is enhancing 802.11 to make it suitable for the vehicular environment. In addition, several industrial consortia have been founded, such as the 'Car-2-Car Communication Consortium' in Europe. Problems on the networking layer include the routing of messages for certain applications where many cars are involved over a large area. The idea here is to use the GPS information that cars have about their location for geographic routing, as discussed in Section 2.4.1.

2.4.6 Ambient Networking

One main challenge in a world of ubiquitous networking is how to automatically interconnect different dynamic wireless networks. An example scenario is shown in Figure 2-13, where a personal area network (PAN), a train network and a WLAN hotspot interconnect.

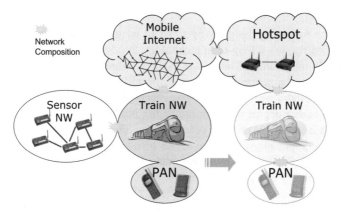

Figure 2-13.   Ambient network scenario and network composition

This challenge is addressed in the European integrated project 'Ambient Networks', which proposes a general framework for dynamic, instant composition of networks.[26] Network composition goes beyond what the Internet and cellular networks provide today. Internet-working will happen not only at the level of basic addressing and routing, but also for additional functions. For instance, mobility handling may be different in different networks. The configuration of control-plane interaction of such networks needs to become autonomic, because it is a very complex process and yet needs to be realized on-the-fly, and moreover transparently to the user.

2.5 Programmable Networks

Designing effective and reliable communication protocols requires a high effort, which is often prolonged by standardization. Hence, introducing innovations often takes a long time. A major approach that leads to flexibility and shortened time-to-market is programmable networks. Programmable network nodes enable the fast deployment of new services by using open interfaces to the network resources. With this technology, new protocols can be installed on the network elements, which use the lower-layer resources for creating new services.

Furthermore, in designing networking protocols and systems, backwards compatibility is often a major issue. However, to be extensible and compatible with future technologies, current designs are often not sufficient. Frequently, the complete firmware of devices has to be updated to allow for new software versions or software bug patches.

Typically, one assumes that more flexibility is needed on the application and middleware layers or higher, and less on the networking layer. Adapting software at run time is now a wide-spread concept in many applications, for example dynamically installing codecs for a video player. Similar situations occur in network protocol design, where only base protocols or mechanisms are deployed and extensions are needed.

For mobile networks, we argue that flexibility in networking is also needed, for several reasons. Mobile networks are very fragile as a result of wireless links and mobility, optimizations pay off more easily. Moreover, we expect considerable innovation in heterogeneous networks with novel air interfaces, which will require adaptation for new technologies.

2.5.1 Adaptability Concepts

We discuss in the following the growing trend to adaptability in communication devices. Flexible adaptation follows a long-term technology trend towards configurable or programmable platforms in many areas. This is due to the progress in the development of semiconductor and software technologies. For instance, for middleware, execution environments like Java are now well established. For lower-layer technologies such as network processors or radio chip sets, a higher degree of programmability is a major current trend.

Our notion of adaptability is broad in order to cover all system aspects, from reconfigurable radio parameters to applications. Because of this broad scope, different forms of adaptability are considered. The supported adaptability could be classified in the following way with respect to configurability and programmability:

- configuration, either before system startup (for example during development or deployment) or at runtime

- parameterization of software as the classical way to introduce flexibility without the need for software updates
- complete software update or exchange, with firmware update of devices as a classical example, which often requires devices or servers to be made unavailable for some period of time
- partial software update in a complex software system, where there are often no open interfaces with the result that most of the expected system behaviour has to be re-evaluated, with possible service interruptions as another consequence
- installation of components in an open platform with an interface design that ensures proper functionality and that supports seamless service evolution
- automatic installation of software that is carried as part of the exchanged data in a communication relationship and that directly influences the exchanged data; this refers to the active networks capsules approach

From this list, we focus on the following open platforms, although other forms are also important in practice. The last item, the capsules approach, offers much higher flexibility regarding the location of the execution of the code, but also possesses very high risks with respect to security.

Although adaptability is an important concept, we should also consider its implications. In many cases, the additional flexibility leads to higher initial cost. However, the price of a system replacement due to inflexibility is significantly higher. The main cost factors, although there are others, are more expensive hardware than initially needed and the development of a reliable programmable environment. In addition, for software, the request for flexibility may lead to more complex designs for the introduction of variation points and indirection layers. On the other hand, programmable physical parameters or flexible software platforms can be used in larger application areas supporting a seamless system evolution that leads to much better economics of scale.

In the following, we introduce a notion of an open platform[27] and show its relationship with the system architecture and cross-layer interfaces. An abstraction layer may consist of several open platforms providing the actual computational behaviour. Each such platform consists of a stable and minimal platform base, plus several platform components, which can be added or removed to address different requirements at different times.

2.5.2 Programmable Network Infrastructure Nodes

A generic architecture for the network elements of a mobile network is shown in Figure 2-14. In this architecture, we consider three platforms, each programmable with configurable components:

- The computing platform serves as a general-purpose platform for processing stateful protocols, such as routing, QoS signalling or connection management.
- The forwarding engine is in the data path of a network node and it connects network interface platforms, for example by a switch matrix. This engine can be implemented as dedicated hardware or as a kernel module of common operating systems. The forwarding engine is programmable for performance-critical tasks, which are performed on a per-packet basis.

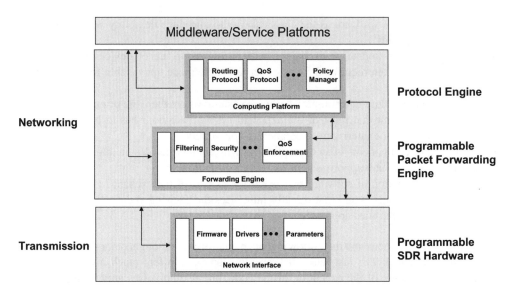

Figure 2-14. Programmable network device architecture (Courtesy of *Journal of Communications and Networks*)

- The network interface platforms are medium specific for different wireless or wired standards. They can be configured or programmed to adapt to new physical layer protocols or for triggering events in higher layers.

From the operational point of view, the main requirements for the platforms are:

- reliability for uninterrupted services, in particular when updating services
- remote management, which is important for the central configuration in large networks
- security with respect to attacks from outside – since we assume that the network is owned by an operator, the main security risks arise from external interfaces

We assume that there is a configuration manager in the middleware abstraction layer, which is able to install new components in the lower layers. Since the configuration will be performed remotely, the distributed processing platform is suitable for this purpose and may also offer transaction services for complex reconfigurations.

2.5.3 Programmable Mobile Terminal Architecture

The main platforms of the terminal-side architecture, as shown in Figure 2-15, are:

- a smart-card networking platform, which provides networking functions such as addressing and authentication
- a smart-card middleware platform, which provides subscriber identities and a highly secure execution environment

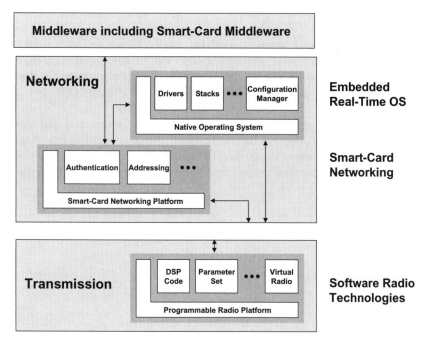

Figure 2-15. Programmable network device architecture (Courtesy of *Journal of Communications and Networks*)

- a programmable radio platform that is designed for one or more radio standards families
- a native operating system platform that provides real time support, needed for protocol stacks and certain critical applications, such as multimedia codecs

The main requirements from operator and manufacturer point of view are:

- survivability, for example robustness to mis-configurations, failures or misuse
- security with respect to end-user, operator and manufacturer requirements. For instance, the manufacturer may be liable if the device does not behave correctly with respect to radio emissions. The operator is mainly interested in reliable services and billing. The user wants to be assured of the device integrity and service availability
- mass-market optimization of hardware and software platforms balanced with time-to market and flexibility or upgrade requirements

A prominent technology for programmable networks is active networking, where the network nodes are programmed to perform customized operations on the packets that pass through them. Active-network technology typically provides an execution environment that is operating-system and hardware independent. There are several approaches to active networks; see, for example, Decasper[28] and Wetherall.[29] Some approaches aim for the high performance needed for flexible, per-packet processing, while others only aim at flexibility in the control path.

2.6 Summary

Wireless communication technologies are evolving rapidly, allowing them to spread to more and more application areas. Future cellular networks will be able to interface to, and even run on top of, several of these diverse technologies. Realizing this extension of the communication domain requires supporting technologies such as efficient QoS and handover mechanisms.

In this chapter we have given an overview of the wireless technologies that are relevant for 4G networks and beyond. Some of them are currently being standardized and (for example WLAN, WiMax and mesh networks) are directly being considered for use in 4G, while others (such as wearable computing and programmable networks) are likely to be further in the future.

3

Mobile Service Systems

Wolfgang Kellerer

Successful next-generation mobile communication systems are expected to focus on the users and on the applications. The service platforms offer the capabilities for innovative applications. Ubiquitous services platforms offer new opportunities for service provisioning.

This chapter provides insights into state-of-the-art and emerging service platforms supporting service provisioning in mobile communications environments. We are not only focusing on cellular systems here, but extendng our focus, as in the previous chapter, to ubiquitous communications environments. This includes service platforms that allow services to be used on various heterogeneous devices and service platforms that are highly distributed such as Peer-to-Peer service platforms. We regard such ubiquitous communications as one of the major future trends that will extend an operator's service environment.

We begin this chapter with an introduction to the topic of service platforms and give an overview over existing platforms including IP-based platforms such as IP Multimedia Subsystem (IMS) and non IP-based platforms such as Customized Application for Mobile network Enhanced Logic (CAMEL), as well as open APIs such as Open System Access (OSA) and mobile Internet platforms such as Wireless Application Protocol (WAP) and i-mode.

Section 3.2 discusses challenges and capabilities considered for next-generation service platforms. The support of ubiquitous services that extend a traditional operator's service platform towards sensor and Peer-to-Peer systems is one of the emerging challenges. Future service capabilities and service features are only briefly outlined because most of them, for example personalization and context awareness, are further discussed in more detail in following chapters.

To illustrate ubiquitous services a concrete system example is given in Section 3.3. In particular, the concept of seamless service continuity in a heterogeneous multi-device environment is explained and a system realization is described as a detailed example, which is based on the Session Initiation Protocol commonly used for multimedia signaling in IP-based service platforms.

Towards 4G Technologies. Edited by Hendrik Berndt.
© 2008 John Wiley & Sons, Ltd.

3.1 Service Platforms at a Glance

Since the time when the analogue telephone service was available world-wide, the landscape of communications services has changed dramatically. There was initially no separate service platform, because a communication system as a whole was just designed for a single service, while today service platforms are faced with manifold requirements for serving an ever-growing application landscape. Providing an abstraction layer for heterogeneous underlying communication systems, service platforms should support different kinds of services provided to the users. The convergence of Internet concepts and telecommunications systems opens up not only new possibilities, but also demands for reliability and openness at the same time. Already today, services are not solely provided by the operators. And the role of so-called third-party providers is further emerging, using operator platforms for service provisioning. Moreover, the interoperability between operators should not be limited to voice roaming alone, but include all kinds of services accessible across the world.

In order to understand the motivations behind different approaches for service platform architectures, in the following we provide an overview of existing service platforms, starting from basic telecommunications standards such as Intelligent Network (IN) to IP Multimedia Subsystem (IMS) and mobile Internet.

3.1.1 Mobile Services and their Support Platforms

Mobile services such as mobile telephony impact our life enormously. However, hardly any term in communications and information industry is used as much in such contexts as the word *service*. Thus, before we take a deeper look into service platforms and their respective architectures, let us provide some definitions that we use throughout this chapter.

In an information and communication system, a *service* is an entity controlling a set of capabilities offered to a user by a provider.[1] It fulfils a special purpose for the user, abstracting from the pure system capabilities. A service is provided to the user via dedicated interfaces. A service could also be the user of another service. In telecommunications a service is an entity that realizes the exchange of information in a specific way for the user. Here, basic elements are connection setup and release as well as adding and removing users, media or connections. Examples of services are videoconferencing, call forwarding, instant messaging and email.

Mobile services use mobile communication systems for the exchange of information, in which at least one hop of the communication path is supported by a wireless link.

Applications implement services that are presented to the users in an application-specific context by providing a particular user interface or application-specific information. In mobile Internet, for example, the WAP (Wireless Application Protocol) system provides the service of Internet access through a wireless link with the WAP browser initialized, the service provider's portal address being the application running on the mobile device to access this service. The data on the web sites accessible through WAP can also be regarded as mobile applications.

The *service platform* describes the part of the system that supports service components. Figure 3-1 shows typical service components, i.e. the features of a service platform, which are:

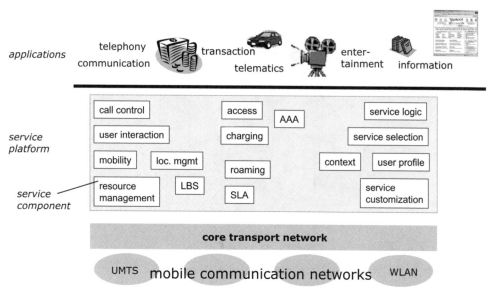

Figure 3-1. Features of the service platform

- control features: call control, session control, user interaction, resource control, Quality of Service (QoS) control
- access features: authorization, authentication, accounting (AAA), discovery, selection, charging
- value-added features: location, context, profile
- mobility support: roaming, service level agreements (SLA)

together with service logic for particular services combining the features.

The following scenarios illustrate the use of services in mobile communication systems.

- **Personalization:** Travelling always means having to make arrangements for many things. With personalized services a user is relieved of selecting and coordinating several tasks at the same time. For example, a business traveller in a rental car heading to her meeting does not need to worry about entering meeting time and destination into the navigation system or adjust information services to the in-car display. Furthermore, the route takes into account a stop at an ATM at a sub-branch of her bank. The reservation for her business dinner with customers is made automatically, even taking the preferences of the customers into consideration.
- **Ubiquitous Service Mobility:** A user who is participating in a video conference call via his mobile phone while travelling arrives at his partner's office. Immediately, his location is registered by the environment and available communication equipment is discovered and presented to him. In order to enjoy a better quality for the videoconference, the user decides to transfer the ongoing session's audio stream to the stereo system and the video to a TV screen in the room.
- **Ubiquitous Peer-to-Peer Services:** For a user who wants to sell opera tickets just before the beginning of the performance, a mobile Peer-to-Peer application running on her device

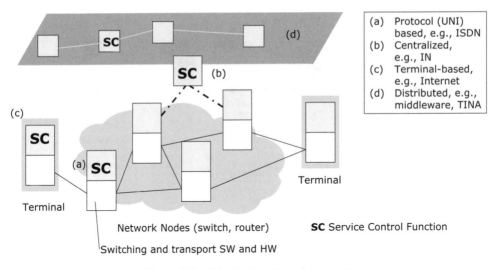

Figure 3-2. Distribution of service control

will be of great help. She describes her offer and interested parties can search for tickets in a Peer-to-Peer manner without having to run around with a signboard. The negotiation and the contract enforcement (payment) is also supported by the system.

In a networked environment a system consists of distributed entities, such as terminals, network routers and network servers. Looking at the different concepts behind service platform architectures we can distinguish several architectural principles along with the location of the (main) control functionality, as illustrated in Figure 3-2. Early multi-service platforms, for example Integrated Services Digital Network (ISDN), have the services integrated in the protocol running at the user–network interface (UNI). This has the drawback that for the introduction of new services all devices and switches have to be updated. Centralized service platform architectures (such as Intelligent Network, IN) have a central server in the network as the main platform hosting the services, which can be more easily created and removed. However, scalability is a problem of these platforms. Internet services are usually provided in an end-to-end manner between user terminals or user terminals and servers. Peer-to-Peer services are the extreme case where the service platform only consists of the user devices. In the mid 1990s, architectures such as Telecommunications Information Networking Architecture (TINA) evolved, hiding the underlying system entities by having a middleware platform interconnecting service platform components running on different entities. Today's architectures are a mixture of these concepts.

3.1.2 Telecommunication Service Platforms

As already mentioned, service platforms evolved from monolithic single-service systems to multi-service platforms as illustrated in Figure 3-3. Let us start our walk through the evolution by considering fixed network service platforms.

From <u>Single Service Networks</u> to a <u>Multi-Service Network</u>

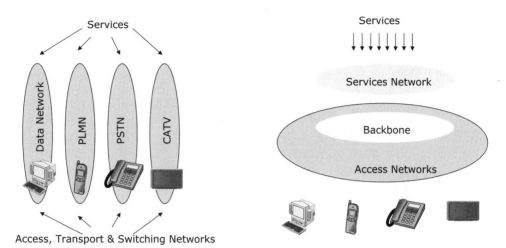

Figure 3-3. Evolution of service architectures: Public Land Mobile Network (PLMN), Public Switched Telephone Network (PSTN) and Cable Television Network (CATV)

In Integrated Services Digital Network (ISDN) systems, service control is included in the layer 3 protocol of the D-channel (Q.931).[2] It includes basic call control for basic services such as voice, fax and data over circuit-switched links and the initiation of ISDN features. Q.931 realizes a decentralized service control, since the protocol is running on all switching nodes. This implies that, for the introduction of new services, the protocol software in all nodes has to be updated. This approach is based on the concept of a closed network where all resources are owned and controlled by one operator.

The Intelligent Network (IN) is an architectural concept for realizing the execution of supplementary services in a distributed, protocol-based environment like ISDN.[3,4] In traditional systems, the service control is directly attached to the switching node. As shown in Figure 3-4, the IN concept describes a central control server, the service control point (SCP), which interacts with a separate signalling protocol with all relevant switching nodes. Usually, the signalling system number 7 (SS#7) is used for transporting the signalling messages separately from the basic call processing and information transport. The service signalling protocol messages are defined by the Intelligent Network Application Part (INAP), which is an optional part of SS#7. In this way the INAP message set determines the capabilities of the service control. The IN is visible in many service platform architectures including mobile systems, as we see in the next section. The main motivation for the introduction of the IN for operators was so that they could become independent of the vendors of switching systems for the introduction of new services. Service provisioning by third parties was not a target.

Distributed Processing Environments (DPE), such as those based on a middleware like CORBA,[5] support the development of distributed systems where system components can be spread over heterogeneous platforms without adaptation of the application. DPEs include

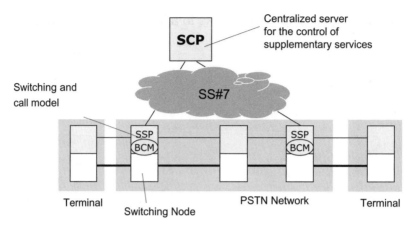

SCP Service Control Point
SS#7 Signaling System No. 7
SSP Service Switching Point
BCM Basic Call Model

Figure 3-4.   Intelligent Network

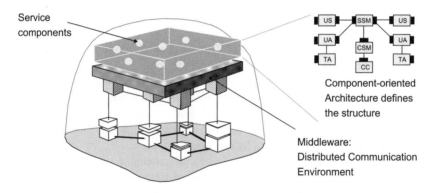

Figure 3-5.   TINA overall architecture

mechanisms for the collaboration of the systems components to exchange necessary information, for example via remote procedure calls.

The Telecommunication Information Networking Architecture (TINA)[6] defines an advanced service, network and management architecture for multimedia information and communication services based on an object-oriented approach (see Figure 3-5). TINA system components communicate via the TINA DPE. In this way the TINA service platform architecture is merely described by the interfaces of the components.

The most important TINA concept is that of a session, i.e. the modelling of the service control functions by interoperating sessions. TINA distinguishes between access sessions, service sessions and communication sessions. Each session involves relevant server functionalities to perform session control, so that for example, the access session deals with user authentication and authorization.

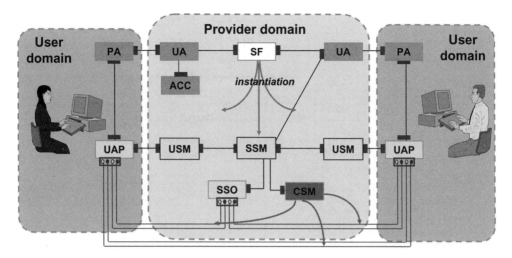

Figure 3-6. TINA service architecture

The TINA service architecture specifies the components that interact in a session and that build the TINA service platform architecture (see Figure 3-6). The access session defines the access of a user to the service platform and its services. A provider agent (PA) component running on the user device connects to the corresponding user agent (UA) on the service-platform provider side. The user agent represents the user in the provider domain and controls accounting (ACC) and service selection. Once the user has selected a service, a service factory (SF) instantiates the respective service control components for this service usage. A service session manager (SSM) as the main control entity, a user session manager (USM) for each participating user and optional special service objects (SSO), representing, for example, content servers, are created. They interface to the service session user application (UAP) on the client side, implementing the user interface. Additional users can be invited through their user agents and join the session with additional components initialized for those users if necessary. The service session manager interfaces to a communication session manager (CSM) that is responsible for the setup of connections between the participating entities as part of the communication session.

TINA has inspired several other service platforms or platform components, for example the separation of access and service sessions in Parlay. However, TINA was never really deployed in a commercial system. The main reasons for this include its early development, which did not consider current IP technologies but was rather based on Asynchronous Transfer Mode (ATM), and the fact that the evolution scenarios that are necessary for interworking with existing systems came late in the TINA specification work.

The request to open up service platforms for other service providers was answered with the Parlay system, which describes a standardized application programming interface (API) to access operator-specific service platform components by third-party providers (see Section 3.1.5). The Parlay API has also been adopted for mobile communication networks referred to as Open Service Architecture (OSA). We will give more details in Section 3.1.5.

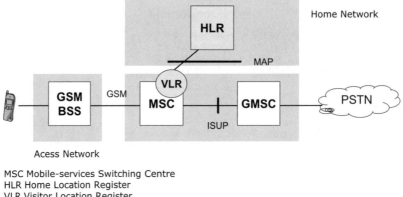

Figure 3-7.   GSM service architecture

3.1.3 Cellular Service Platforms

Influenced by the fixed network telecommunication architectures, mobile network service platform architectures have also evolved from voice-based systems to open service platforms. Moreover, we can also observe how the Internet paradigm of packet-based transmission has been integrated and how this influences the service platform architecture.

The service platform of the European second generation mobile communication system GSM (Global System for Mobile Communication)[7] is mainly represented by the Home Location Register (HLR) in addition to the Mobile-services Switching Center (MSC), which is responsible for basic call control (see Figure 3-7). The HLR can be compared with the SCP in the Intelligent Network architecture, hosting user data such as the current location for mobility control and additional supplementary services. The HLR services are focused on user-related services such as call forwarding on busy that are invoked by triggers in the basic call control. For global supplementary services such as number translation another central server has been introduced with CAMEL.

CAMEL (Customized Applications for Mobile Network Enhanced Logic) represents a further IN-based overlay in the GSM system.[8] In addition to the service control point functionalities in the GSM HLRs focusing on personal services, the CAMEL architecture provides additional service control points, called gsmSCF, to cover global services (for example number translation for free call services, virtual private networks and short message service – SMS) separately from the basic call control in the switching nodes. Figure 3-8 illustrates the CAMEL architecture in the GSM system. The gsmSSF hosts the Service Switching Function detecting triggers for a supplementary service call.

The evolution from second to third generation mobile systems was, from a service platform point of view, characterized by the introduction of packet-based communication. The General Packet Radio Service (GPRS),[7] which is often said to be a 2.5 generation, introduces a new domain to the so far circuit-switched transport. New components such as SGSN (Service GPRS Support Node) realize IP-based communication within the mobile communication

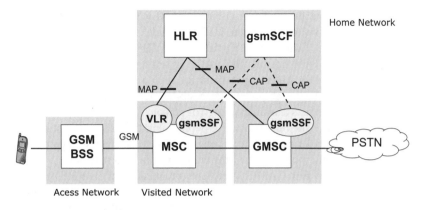

SCF Service Control Function
SSF Service Switching Function
CAP CAMEL Applic. Part of SS#7

Figure 3-8. CAMEL

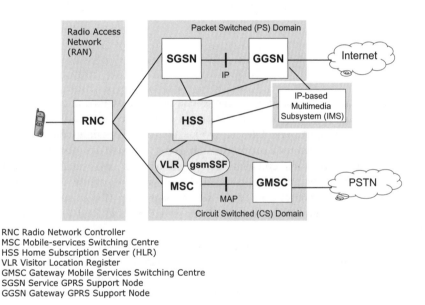

RNC Radio Network Controller
MSC Mobile-services Switching Centre
HSS Home Subscription Server (HLR)
VLR Visitor Location Register
GMSC Gateway Mobile Services Switching Centre
SGSN Service GPRS Support Node
GGSN Gateway GPRS Support Node

Figure 3-9. UMTS service architecture

system and the GGSN (Gateway GPRS Support Node), the interconnection to the global Internet.

The third generation, represented by UMTS (Universal Mobile Telecommunication System)[9] in Europe includes both domains, offering circuit-switched voice communication and packet-based Internet connection. In addition, the IMS (IP-based Multimedia Subsystem) was introduced to control IP-based multimedia services in UMTS (see Figure 3-9). We will describe the IMS service platform in detail in the following subsection.

Beneath IMS for the packet-switched domain and HLR/CAMEL, which are carried over from GSM to the circuit-switched domain, UMTS includes further service architecture components described by the Virtual Home Environment (VHE).[10] The VHE is a concept for a personal service environment hosting personal user setting (user profile, user preferences, personal services) in UMTS. Main requirement for such a personal service environment as specified by the VHE is the portability and accessibility across networks and between terminals. The VHE concept targets a seamless usage of personal services and profiles of the home network environment in visited networks. For deployment the VHE concept specifies a very generic architecture with service execution environments and data storage points in every mobile system component including the SIM card, terminal and network.

For the circuit-switched part, CAMEL represents a realization of the VHE concepts because CAMEL services are accessible based on the IN technology during roaming. However, this is limited to voice services, which are the target of the CAMEL service platform. Other components required for a realization of the VHE in UMTS are the Mobile Application Execution Environment (MExE) and the UMTS SIM Application Toolkit (USAT) (see Chapter 5). These define an environment for the execution of services in the mobile terminal and in the SIM card, respectively. Respective download mechanisms and security features make these environments reliable platforms for execution of personal services in mobile phones.

3.1.4 IP-based Mobile Service Platforms

Services in the Internet are based on an end-to-end principle, removing complexity from the network servers and allowing a simpler creation of new services and applications based on terminal software only. The Internet service platform is mainly characterized by the control protocol that is used for a service session. Here, a *session* describes a set of meaningful communications between endpoints supported by one or several data connections. The data transport can be tied to the session signalling (for example, HyperText Transfer Protocol, HTTP) or is independent of the session signalling (for example in a streaming session, Real Time Control Protocol (RTCP) for signalling, and Real Time Protocol (RTP) for transport).

For multimedia service control on the Internet two main protocols have been standardized for session signalling. The H.323 protocol suite, standardized by the ITU-T (first release in December 1996), is referred to as a 'standard for real-time videoconferencing over non-guaranteed quality of service LANs'.[11,12] The Session Initiation Protocol (SIP), standardized by the IETF Multiparty Multimedia Session Control Working Group (MMUSIC WG) as a proposed standard first in 1999, is the basic protocol for the IETF multimedia service platform.[13] The two protocols offer similar features and have been designed for a broader usage beyond voice over IP (VoIP). H.323 (together with the standard H.450) is especially focused on telephony supplementary services and smooth interworking with the telephone network to replace private branch exchanges (PBXs) with VoIP. SIP has a broader scope from the beginning, focusing also on non-VoIP services (such as Instant Messaging). SIP defines only elementary protocol messages, which are composed in relation to service features. More on the service architectures of both protocol suites is given by Glasmann.[14] For session signalling on IP, the UMTS standardization body 3GPP has selected SIP as the basis for IMS.

3.1.4.1 Session Initiation Protocol

The architecture of the SIP service platform consists of the basic SIP protocol and its extensions, running on user terminals implemented as SIP user agents or in network servers such as SIP proxy servers. Furthermore, additional services can be implemented on proxy servers or on SIP user agents interfacing to the SIP protocol. Several scripting languages have so far been proposed for SIP service development, including the Call Processing Language (CPL)[15] for services on SIP proxies. In summary, the SIP service platform consists of several levels, as shown in Figure 3-10, which can be used for the realization of services such as Instant Messaging (IM), telephony and telephony supplementary services. The SIP core protocol is the basis for this and supports most service scenarios. SIP extensions are generic features for certain service types. Service scripts in SIP entities further refine SIP protocol logic. It is important to note that SIP is designed as a general-purpose session initiation protocol, not limited to replication of telephone features, and thus all extensions are considered carefully in the respective working groups of the IETF.

As SIP is the basis for IMS and, furthermore, we use SIP in the concrete system example in Section 3.3, we provide more details about the protocol here so as to provide a better understanding of the protocol concept. For the establishment and control of sessions, SIP provides a small set of text-based messages (called methods) that are exchanged in separate transactions between the SIP peer entities (SIP user agent in a user terminal).[13] Methods such as REGISTER, INVITE or BYE specify actions that have to be performed by the participants involved. Each method consists of a request message sent by one participant and a number of response messages. For example, an INVITE request indicates to the recipient that a multimedia session is to be established. If the recipient agrees, a '200 OK' message is the proper response. This has to be acknowledged by an ACK method sent by the original sender, completing the session establishment transaction. The session data path is set up independently

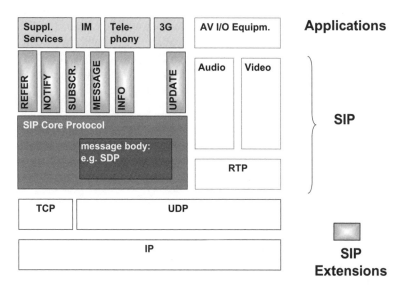

Figure 3-10. SIP service architecture

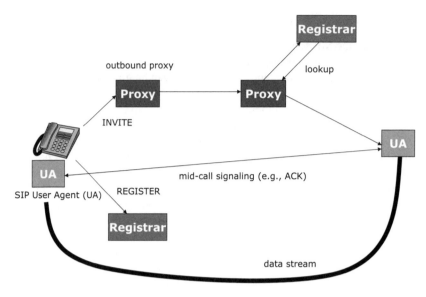

Figure 3-11. SIP trapezoid

from the signalling path, which is fundamental for SIP. In the signalling path, network enti-ties, such as proxy servers or redirect servers that can be traversed by the messages, can be used for support, for example for address resolution. Figure 3-11 shows a typical SIP system setup with the INVITE message traversing several proxy servers and already the ACK message taking an independent path, as does the data stream. This is called the SIP trapezoid.

The session itself is described in two levels. Whereas the SIP protocol methods contain the parties' addresses and protocol processing features as method headers, the description of the media streams that are exchanged between the parties of a multimedia session are defined by another protocol. The IETF suggests the Session Description Protocol (SDP),[16] which in fact is not a protocol, but a structured, text-based media description format that can be carried in the SIP message body. This is an example of a SIP message:

INVITE sip:bob@docomo.com SIP/2.0
Via: SIP/2.0/UDP pc12.company.com
To: Bob<sip:bob@docomo.com>
From: Alice<sip:alice@company.com>
Call-ID: a84b4c76e66710@company.com
CSeq: 1 INVITE
Contact:<sip:alice@company.com>
Content-Type: application/sdp
Content-Length: 153
v = 0
o = alice 53655765 IN IP4 pc33.company.com
s = Session SDP
t = 0 0

c = IN IP4 pc33.company.com/127
m = audio 3456 RTP/AVP 0
m = video 51372 RTP/AVP 31
m = application 32416 udp wb

The upper part shows an example of the SIP INVITE methods with typical headers detailing the request message. The first line indicates the method sent to Bob in the domain docomo. com based on SIP version 2.0. This request has to traverse the server pc12 of the originating domain (Via: header). The To: and From: headers indicate receiver and originator of the message. An optional Contact: header can be used to describe the responsible session controller in case of, for example, third-party call set up. The Call-ID: gives a unique session identifier and the CSeq marks the sequence of messages within the request message flow. Finally, the Content-Type: and Content-Length: headers describe the attached session description, which is SDP in this case.

The SDP description starts with the version number (v), the originator's name, a session identifier and the IP address of her host (o), the session name (s) and the session schedule (t = 0 0 indicates immediate start and non-bounded duration). The connection data (IP address and port number) of the originator's application are stated in the c line followed by the media streams described by type (audio, video or specific application such as white board/wb), port number and transport protocol (here using RTP with the IETF Audio/Video profile) with a format descriptor (for example 0 for PCM).

3.1.4.2 IP Multimedia Subsystem

3GPP, the standardization organization for the IMT2000 family, has specified the IP Multimedia Subsystem (IMS) as a system architecture supporting multimedia services in a packet based environment.[17] It should allow the operator to create and control new services more flexibly, based on Internet principles. In particular, IMS allows operators to have more control over the service provisioning than just becoming a bit-pipe, providing Internet access via mobile data networks such as GPRS. This is achieved by a correlation of the SIP-based service layer and the transport in the packet-switched domain. Several features reflecting operator requirements such as Quality of Service (QoS) control, charging and subscription management have been added to the SIP protocol. IMS is a part of the packet-oriented domain of IMT2000, which complements the circuit-switched domain. As it is planned in the future that all services will converge to packet-based operation (IP-based), IMS will not be restricted to the core network, as it is now.

Figure 3-12 shows the basic components in the 3GPP IMS specification. The lines show different information paths: separate paths for signalling and data in the circuit-switched domain and a combined path for the packet-switched domain. The dashed line indicates signalling information. On the left-hand side the access network (AN) is depicted with the Radio Network Subsystem (RNS). The packet-switched core network (PS CN) domain and the circuit-switched core network (CS CN) domain are shown in the middle, both interfacing to the Home Subscriber System (HSS), the successor to the GSM HLR. The IMS sub-domain is illustrated on the right-hand side. Basically, the 3GPP architecture is a combination of SIP (for session initiation and mid-session call control) and MEGACO[18] for the control of media

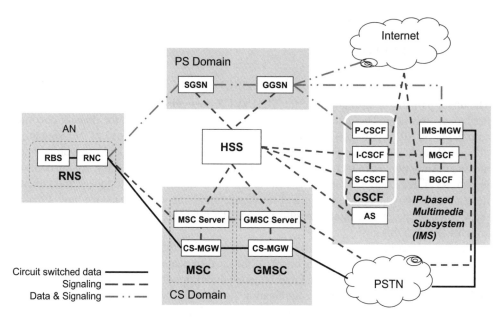

Figure 3-12. IMS architecture

streams through media gateways. The UE is acting as a SIP User Agent, initiating and terminating SIP requests. The three IMS SIP server types are the Proxy Call Session Control Function (P-CSCF), the Interrogating CSCF (I-CSCF) and the Serving CSCF (S-CSCF). The Application Server (AS) can be regarded as a SIP endpoint, providing services such as media streaming. The CSCF servers control the data stream and in particular the interworking with the conventional telephony network through media gateways (IMS MGW) and media gateway control functions (MGCF and BGCF).

The P-CSCF is the first contact point of a roaming IMS user. It acts as a proxy to the user's home network IMS domain. The I-CSCF is the contact point between different operators' IMS domains. It hides the configuration, capacity and topology of an IMS system from outside and performs the authentication and authorization of user requests to the home domain. The I-CSCF knows the user's HSS and the respective S-CSCF for service execution. The S-CSCF performs the session control and handles the session states. It also acts as a registrar server being connected to the HSS for subscription and profile maintenance. In this way, the P-CSCF could be regarded as an SIP Outbound Proxy, being the first termination point of a SIP request, routing the request further to the destination (DNS lookup). I-CSCF and S-CSCF represent the network servers supporting user registration and service access. For setup and during the session the S-CSCF acts as an SIP proxy server relaying messages between the endpoints, which could also be application servers.

It should be noted that the CSCF modules are not implemented as real SIP proxies but as so-called Back-to-Back User Agents that terminate calls on the one side and originate the same call on the other side instead of simply relaying them. Thus the end-to-end signalling model of SIP is no longer valid in an IMS environment. This is, moreover, a closed environment, because additional signalling messages or SIP headers (which are allowed by the SIP specification) are terminated in the P-CSCF and not proxied further. This enforces a net-centric, rather than user or third-party driven, service creation.

IMS was originally designed for mobile networks and thus standardized by 3GPP. Recently, TISPAN (Telecoms & Internet converged Services & Protocols for Advanced Networks), a standardization body of ETSI focusing on Next Generation Networking, has included IMS in its fixed network architecture. In this way IMS provides support for fixed and mobile networks, paving the way for fixed–mobile convergence.

3.1.5 Open APIs

The need to open operators' networks and service platforms to other, so-called third-party, providers was first addressed by the Parlay API.[19] This API, which is standardized by the operator- and manufacturer-driven Parlay consortium, enables third-party applications to make use of network functionality through a set of open, standardized interfaces. For mobile communication systems the Parlay specification has been adopted by the 3GPP as Open Service Access (OSA) and is currently being jointly developed further.

Figure 3-13 illustrates the OSA architecture specifying the OSA API of the OSA gateway. This API allows an application hosted by an application server to access service capability servers that are providing specific service and network control functionality. The service capability servers map the methods available at the OSA API to network-specific protocols and thus shield network-specific details from the application. Table 3-1 gives an overview of the most important service capability functions that have been standardized for OSA.

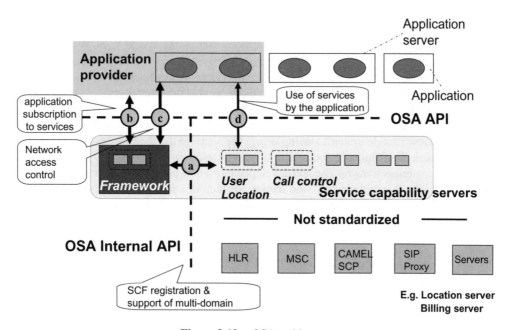

Figure 3-13. OSA architecture

Table 3-1. OSA service capability functions

CAPABILITY SERVER	DESCRIPTION
Call control	Setup and control of basic calls and multimedia conference calls
User interaction	Obtain information from the end-user, play announcements, send short text messages
User location/User status	Obtain location and status information
Terminal capabilities	Obtain the capabilities of an end-user terminal
Data session control	Control of data sessions
Generic messaging	Access to mailboxes
Connectivity management	Provisioned QoS
Account management	Access end-user accounts
Content-based charging	Charge end-users for use of applications/data

In addition to the interfaces allowing access to service capability functions (d in Figure 3-13), OSA has standardized two further interface classes. The framework interface is responsible for controlling the access of an application to the OSA gateway before using service capability functions (b and c in Figure 3-13). An internal OSA API (a in Figure 3-13) describes registering of service capability functions and authorizing access of an application to these.

Whereas the OSA API is a first step to opening a mobile system to third-party providers, its usage is still complex and requires more sophisticated knowledge of telecommunication concepts than is required for service creation in the Internet.

3.1.6 Mobile Internet

The success story of the Internet and its applications such as the World Wide Web (WWW) is based on its ease of use by customers through browsers and the easy creation of new applications, not only by communication service providers but any kind of business or private user. It is obvious that operators want to repeat the success of the Internet on mobile networks. In order to provide a WWW-like information service to mobile users, different service platforms have emerged. In this section, we will discuss the Wireless Application Protocol (WAP) and the Japanese i-mode system. Having nearly the same goals, their success has been different. Whereas WAP initially focused on a generic platform infrastructure optimized for low-data-rate wireless transmission and developed into a new protocol architecture, the scope of i-mode was on the applications, with the introduction of a new business model, in which third-party providers can easily provide new applications to mobile users. Let us have a closer look on the two service platforms.

3.1.6.1 Wireless Application Protocol (WAP)

The mobile Internet service platform known as the Wireless Application Protocol (WAP)[20] was designed as an IP-based, WWW-like information service for mobile users. The main focus was put on the infrastructure, resulting in an optimization of the entire Internet protocol

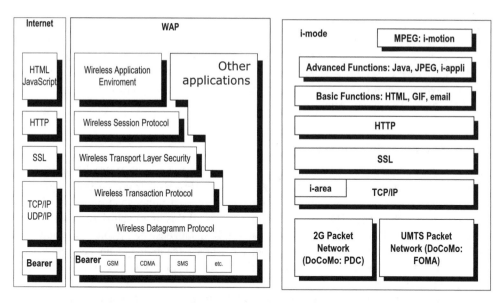

Figure 3-14. Comparison of WAP and i-mode protocol stacks

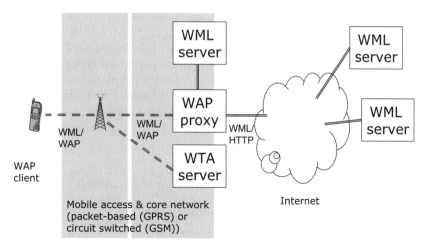

Figure 3-15. WAP service architecture

stack for the wireless connection and the introduction of the Wireless Markup Language (WML), a new, optimized markup language different from HTML to adapt the information to the limited display capabilities of the mobile terminal. Figure 3-14 (left-hand side) illustrates the WAP protocol stack in comparison with Internet protocols used for web browsing. WAP is standardized by the Open Mobile Alliance (OMA, formerly WAP Forum) an industry consortium.

Figure 3-15 gives an overview over the WAP service architecture. The WAP architecture basically consists of a WAP client, a WAP proxy and a WML server. In the mobile network,

HTML over HTTP/TCP/IP is replaced by WML over the WAP protocol stack up to the WAP server. WAP supports circuit-switched bearers such as GSM data channel and packet-based bearers such as GPRS. The WAP proxy acts as a gateway to WML servers that are either in the mobile network operator's packet network domain or accessible via the Internet. In addition, an environment for application development for mobile communication devices is provided (Wireless Telephony Application, WTA).

3.1.6.2 NTT DoCoMo's i-mode

The objective behind the development of i-mode was the same as with WAP: introduction of an Internet-like information service for mobile users.[21] However, from the very beginning, the focus was on the services and applications to be offered rather than on developing an ideal transport technology. The success of i-mode proved this approach right. Today, there are more than 46 million i-mode users in Japan (50% of Japanese cell phone users and 91% of NTT DoCoMo's customers). In addition, the i-mode technology and concept is also exported worldwide. i-mode is deployed by various operators in 15 countries worldwide (June 2006; source NTT DoCoMo).

The service architecture of i-mode (see Figure 3-16) is very similar to that of WAP, showing some simplification in the protocols and additional features such as built-in email. The i-mode client is connected via a packet-based network to the i-mode server hosting the i-mode portal, which lists all official i-mode content providers and allows subscription to their services, and the i-mode gateway to the Internet for information and for email. In addition to the official i-mode servers that have to be certified by the operator, an unlimited number of non-official servers are possible, simply offering i-mode conformant pages over HTTP.

In comparison to WAP, which has not yet achieved much commercial importance, i-mode is characterized by a number of factors, which led to its commercial success. For transport, i-mode has used a packet-based network, which is similar to GPRS, over a wireless link from the beginning. The introduction of i-mode was very much bound to the introduction of NTT

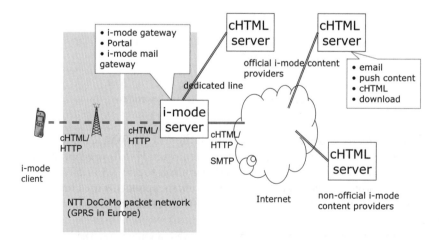

Figure 3-16.   i-mode service architecture

DoCoMo's packet-based network. While WAP runs over various transport bearers including GSM data channel and even SMS, this meant an additional threat for the i-mode operator at that time. However, as a result of this, the i-mode service received better customer acceptance from the beginning, because of the higher data rate and the possibility of always being on, i.e. packet-based accounting. i-mode is much closer to Internet standards than WAP. Instead of the highly optimized, but difficult to learn, WML, i-mode pages are described with cHTML, which is just a subset of HTML disallowing some features for performance reasons. In addition, i-mode uses standard HTTP/TCP/IP over the wireless link. In this way, performance is not improved very much by optimized protocols. Rather, i-mode performance is controlled through the content. For certified i-mode pages, the layout and size per page is given by the operator and strictly controlled. This leads to guaranteed short download times, very much improving customer acceptance.

So far we have mainly discussed technology factors. However, one of the main success factors of i-mode is the innovative business model (Figure 3-17), which can be considered as implementing a new paradigm for the provisioning of mobile services. It not only introduces the role of third-party content providers, but at the same time aims to create a win–win situation between the mobile operator and the content providers. i-mode-certified content providers offer their content through the i-mode portal to the operator's customers against a service fee. This service fee is collected by the operator and forwarded to the content provider. Only a very small portion is kept by the operator (9% in Japan). The operator mainly benefits from the packet transport charges and from the customer binding as content providers are encouraged to provide new innovative services because they can rely on the operator for the management, charging, etc. The operator also controls the number and type of i-mode certified sites (listed in the portal) and not only their technical compatibility. To be accepted, new content providers have to offer different services from those already available. In this way, the user is offered a balanced portfolio of services, rather than being overwhelmed with masses of similar offers.

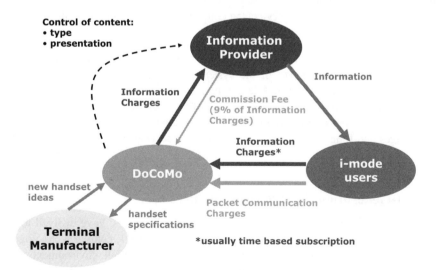

Figure 3-17.   i-mode business model

The relationship with mobile phone manufacturers is also different from the relationship in WAP introduction. The WAP standard has to deal with many different device configurations and, in practice, a WAP page had to be tested against all main mobile phone browsers to guarantee its proper display. NTT DoCoMo's relationship with mobile phone manufacturers is much closer than that in Europe, for example, exactly specifying device capabilities together with the manufacturer and branding devices exclusively, which guarantees the technical compatibility. This branding can also be seen in Europe nowadays.

Today, i-mode is not only available outside Japan, but the business model has also been applied by other operators. Moreover, the convergence of WAP and i-mode is under way as a result of the introduction of XHTML support for mobile handsets supporting cHTML and WML. Furthermore, WAP 2.0 is using HTTP and SSL replacing WSP, WDP und WTLS for a smoother Internet convergence.[20]

In summary, the examples of WAP and i-mode for mobile Internet have shown that it needs more than technology for the introduction of successful services in mobile communication and beyond. Alliances with content providers and, in particular, an in-depth consideration of user acceptance are regarded as key requirements.

3.1.7 Requirements for Mobile Service Platforms

Based on the above discussions, we can summarize the main requirements for service platforms (not only mobile service platforms) in the following:

- Provide interfaces to service and application developers, abstracting from network details and supporting value-added features such as user support, access, mobility or location.
- Performance and reliability should be similar to what users experience from fixed networks.
- Support access to the Internet and its services either directly or using the Internet infrastructure to build internet compatible packet-based services.
- Provide interfaces to external service and content providers in order to enhance the service and applications portfolio.

As we have seen from the evolution presented above, the service platform has emerged from an implicitly contained part of the system to become an explicit component that is responsible for the success of a mobile communication system. It also seems unlikely that the evolution of the service platform will stop where it currently is and it will evolve like other system components for future mobile service provisioning. We will describe the emerging challenges and functionalities for next-generation service architectures in the following section.

3.2 Next-Generation Service Architectures

Mobile communications, providing a growing variety of services, have evolved to become an integral part of our daily life. As we have seen in Section 3.1, cellular technologies have been enhanced by Internet technologies in order to repeat in mobile environments the enormous success of Internet services. A recent trend as been towards ubiquitous computing, with the introduction of interaction with the environment based on various devices, including sensors and actuators. Furthermore, to enable broadband multimedia, the industry is pushing a variety of new standards that allow high-data-rate mobile multimedia applications.

In such diverse world, the success of the next-generation mobile communication systems will depend on services and applications that will be provided. Future service platforms are expected to integrate the different paradigms providing open interfaces to service and application providers. New software technologies such as web services and semantic web are likely to play a major role in future service provisioning. New paradigms, adding additional requirements to those stated in Section 3.1.7, are emerging. For example, customer acceptance is considered to be widely increased by tailoring services and applications to actual user needs, to their preferences and to the user's context. Another example is peer-to-peer services, where mobile users directly interact with each other without central control. A well engineered next-generation service platform should provide all means to allow innovative services to be created, deployed and managed in order to address customer and provider needs. For example, third-party interfaces will allow chaining of expertise in service provisioning. In addition, semantic technologies may help to structure contextual knowledge about the user's environment.

3.2.1 Challenges

Based on the above considerations, we outline emerging observations from different viewpoints that indicate the challenges for next-generation systems. From a service life-cycle viewpoint, we can predict the extension of the service life cycle towards more customer focus. From a networking viewpoint, new emerging communication systems, extending beyond cellular networks and including, for example, wireless LAN or sensor networks, have to be considered for service provisioning within what is described by the term *ubiquitous communications*. The opening up of systems, leading to new role models for the players and thus further competition, has to be considered when one is looking for successful business models.

The well known service life cycle consists of service creation, deployment, usage and termination. In terms of customer focus, we should look more closely at service usage, which is the part of the service life cycle that is seen by the customer. Here, an increasingly user-centric view leads to an extension of the simple service selection, execution and termination pattern. Figure 3-18 shows how such an extended service usage life cycle could look. Services can be selected by the users and be based on the user's current context. Thus, information processing is preceding the selection phase. Furthermore, with systems being opened up to multiple providers, there will no longer be just one service available for each service type, but a huge variety, from which the user has to make a selection. A future service provisioning system should help the user to discover suitable services from which a selection can be made. In addition, services can be formed of other services in real time. To suit the preferences and the context of the user, services will have to be adapted and configured before they are executed. Continuous adaptation may also be necessary during execution through to service termination. The extended view on the service-usage life cycle that is shown in Figure 3-18 should serve as an example, although the order of the phases is not necessarily fixed.

From a communications perspective, the extension of the service platform towards ubiquitous environments such as wireless data networks or sensor networks will have to be taken into account for future mobile systems.[22] The definition of ubiquitous services presented

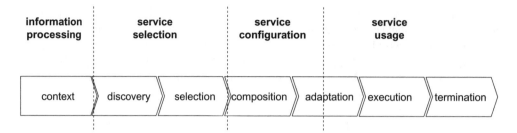

Figure 3-18. Extension of the service usage life cycle

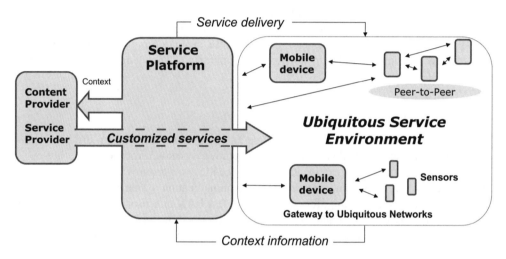

Figure 3-19. Extension of the service platform to ubiquitous environments

here takes a rather broad view. Ubiquitous services are services that make use of communication environments that are not part of existing cellular system. In this way ubiquitous services include mobile services that use environmental information (for example context information gathered from sensor networks), that are accessible and usable in heterogeneous environments (effectively anywhere) and that may use non-cellular network infrastructure (such as peer-to-peer services). Figure 3-19 illustrates this extension of traditional service platforms. The reach of the mobile operator's service platform could be extended through the incorporation of, and interaction with, for example, sensor systems and peer-to-peer systems. The mobile phone is considered to be taking a key role as a gateway, bridging between the different networks. Emerging environments could not only be used for service delivery to devices but also to provide valuable information about the environment, which can be used for service customization in the service platform or by third-party service providers.

As indicated on the left-hand side of Figure 3-19, the roles in mobile service provisioning are already blurring as a result of new market entrants such as content providers and wireless access data network providers. As with number portability and Internet gateway selection, traditional mechanisms for customer binding are vanishing and operators have to develop new ways to retain customers and to enhance customer relationships. One way would be the increased customer focus achieved by offering personalized services, while another would be the opportunity to customize services for third-party providers, making use of the rich information that operators already have about their customers. Such service personalization features used by third-party service providers for their customers could also include simple service access and the operator's role as a trusted party for accounting and service delivery.

Based on the above considerations in service provisioning, networking and role models, new requirements for mobile service platforms include the following:

- focus on the individual user and user groups (personalization, communities)
- consideration of the situation in which an application is running and the surrounding environment (context awareness)
- consideration of all ubiquitously available devices surrounding a user and support for their utilization
- consideration of emerging communication technologies (sensor networks, peer-to-peer communication)
- allowing seamless services in heterogeneous environments and across operator domains
- supporting an increasing diversity of rich services and applications (mobile multimedia)
- simplicity of service access, usage and accounting
- business models that integrate operators, content providers and other third-party service component providers
- customer binding through trust and reputation

Service platform technologies realizing the above requirements originate from various disciplines in computer science, electrical engineering and beyond. As we will see in the following sections and chapters, only a convergence of technologies from different areas can provide the solutions needed for innovation.

3.2.2 Common Service Features

This section briefly lists the most important service features of emerging service platforms. Some of them will be discussed in more detail in Chapter 5, which deals with mobile middleware and includes a description of middleware functions for mobile systems that support the emerging capabilities described above. These middleware functions can be seen as being on a higher level than the common service features or the middleware components for service support listed immediately below.

In addition to conventional middleware features for service support, such as a repository data base, performance management and transaction monitoring, common service features include:

- discovery and advertisement so as to dynamically publish and search for information describing services across domains
- transformation to support the adaptation of content
- membership and management of groups
- negotiation to allow users to discuss an agreement with a service provider

- legal and contract services to oversee the fulfilment of agreements
- a domain schema model providing a conceptual model for structured data to be used for the handling of, and reasoning about, context information
- composition to allow services to be built dynamically out of other services or service components, including services from other domains

3.2.3 Ubiquitous Services Features

According to the definition of ubiquitous services in Section 3.2.1, a ubiquitous service is characterized by one or more of the following:

- using environmental information (for example context information gathered from sensor networks)
- being accessible and usable in heterogeneous environments (at any time and from anywhere)
- relying on non-cellular network infrastructure (such as ad hoc peer-to-peer services)

The provisioning of ubiquitous services requires additional capabilities in the service platform. Gateway functionality has to be provided that supports the search for, and the acquisition of, information in ubiquitous environments. Furthermore, not only sensing, but also the activation of services on ubiquitous devices, has to be supported. Seamless access and usage have already been addressed in the description of the seamless access feature in Section 3.2.2. Here, we can further add that seamless ubiquitous services have to be aware of the environmental context of a user in order to minimize user interaction. In Section 3.3 we will describe a service-platform component that realizes seamless service mobility. In addition, when considering ubiquitous services, we also have to take into account service platforms that are not based on any server equipment, but use the customer equipment for service provisioning. Whereas this appears to be a huge threat to traditional operators, it also provides advantages in terms of infrastructure savings and, from a service provisioning point of view, widens the scope of services towards user-defined services. We will discuss peer-to-peer service platforms, and in particular mobile peer-to-peer platforms, in Chapter 4.

3.3 Ubiquitous Services Example: Session Mobility

In a future mobile communications scenario, the user may have many communication devices (mobile phone, IP-phone, wearable device, etc.) with different kinds of access networks (such as Wireless Local Area Network (WLAN), fixed DSL network and cellular network). Such an environment is usually called a ubiquitous environment and is illustrated in Figure 3-20.

In this section we illustrate with a concrete system example the ubiquitous services that are emerging in such a heterogeneous multi-device environment. In particular, the concept of seamless service access and service continuity is explained by focusing on mobility support on the application layer. In addition, a system realization is described as a detailed example, based on the Session Initiation Protocol commonly used for multimedia signalling on IP-based service platforms.

Before we get to the system example, we first introduce several categories of mobility. In addition to well known terminal mobility, which is merely a task of the network layer to support

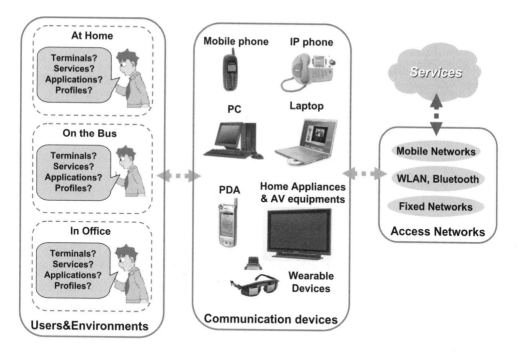

Figure 3-20. Ubiquitous and heterogeneous environments

handover and roaming, other types of mobility are receiving increasing attention in the upcoming user- and service-centric world. As the state of the art shows, the application layer is able to provide mobility support beyond terminal mobility and examples include SIP,[24] ICEBERG,[25] TAPAS,[26] SSE[27] and BSPM.[28] We consider as a basic advantage for mobility control using application-layer signalling protocols such as SIP, that they provide a unique user address, for example the SIP URI (Universal Resource Identifier), which is independent of the network (IP) address. Network-layer mobility and service-layer mobility still have to work together to provide the user with all possible degrees of freedom in mobile communications.

3.3.1 Categories of Mobility

Originally, two kinds of mobility, namely terminal and personal mobility, were recognized. To support human needs and to make life more comfortable, three types are now receiving increasing attention, namely service, profile and session mobility.

3.3.1.1 Terminal Mobility

Terminal mobility is the ability of a terminal to change its location while maintaining active communication (one terminal – several networks). Terminal mobility can be categorized into two types, horizontal handover and vertical handover. With horizontal handover the terminal moves between cells of the same access network type. With vertical handover the terminal

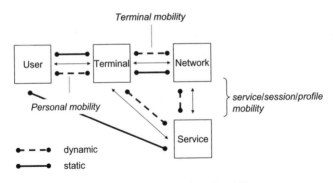

Figure 3-21.   Categories of mobility

moves between different access network types, for example from UMTS to WLAN. The
dynamic relationship between terminals and networks is illustrated in Figure 3-21.

3.3.1.2 Personal Mobility

Personal mobility is the ability of end users to originate and receive calls and access telecommunication services on any terminal at any location, as well as the ability of the network to
identify end users as they move (one number/address – several terminals). Personal mobility
is based on the use of a unique personal identity (for example a personal number or a personal
address). As shown in Figure 3.22, the mapping of user and terminal(s) is considered to be
dynamic.

3.3.1.3 Service Mobility

Originally, the concept of service mobility relies on the Virtual Home Environment (VHE)
concept, which allows users to use and maintain their services independently of the location
of the user and the terminal, for example to have a personal address book available independent of the device type or the network operator. In general, service mobility means the portability of the service environment, or parts of it such as appearance, features and user
interface, across networks and terminals (one service – several terminals and domains). As
shown in Figure 3-21, the dynamic relationship between terminals, network domains and
services not only poses the question of how to move services but also that of how to discover
services and devices. For example, a user changing to a new location may want to initiate
his or her services from a more powerful device that is available at the new location (device
discovery).

3.3.1.4 Profile Mobility

Profile mobility is a sub-problem of service mobility. It describes the portability of personalized information and features of a service environment across terminals and networks (and
even across service environments). This means that profile mobility is the ability of users to
access user profiles and user preferences from ubiquitous locations on any device.[29]

In current solutions, where profiles are made available to various services, these profiles are usually stored locally, for example in the user's terminal (GSM SIM card, PDA or personal computer). The profiles are then synchronized manually among the services. There also exist solutions where profiles for specific services are stored in the local network or in centralized servers, but these solutions are usually proprietary or focused on a particular network. General solutions spanning different networks, services and terminals are missing. As a result, today's users have many personal databases that contain the same information. For example, users store their telephone numbers in their cellular phone, their SIM card, their office phone, their PDA and their Web browser's address book.

3.3.1.5 Session Mobility

Session mobility is the ability of the user to move an active session to another terminal, while keeping the session active. It is a sub-problem of service mobility, describing the transfer of an executed service (and maybe also the active service environment). For example, when a mobile user has a video call with his friend on the way home, once he arrives at home he may transfer the call to other terminals having a better capability, such as a bigger screen or better-quality audio.

3.3.2 Service Mobility in a Ubiquitous Service Environment supported by SIP

In the following, to illustrate the functionality and implementation of ubiquitous service support in a common IP-based environment, we describe a SIP-based system architecture providing a context-based transfer of services. In particular, a mobile user with a mobile device entering a new area (for example a company conference room) is able to:

- Dynamically and automatically discover all available heterogeneous (programmable and non-programmable) devices in the user environment, as well as their capabilities (such as supported media types or codecs).
- Customize the discovered devices by transferring the user profile to the device configuration.
- Update the configuration of all other active devices in use when changing the profile on one device.
- Transfer an ongoing real-time multimedia session (say a video call) from the mobile device to one or more devices discovered in the entered area. This also includes the capability to split sessions, for example splitting an audio and video session into two sub-sessions (one for audio and the other for video) and transfer each sub-session to a different device.
- Re-transfer all transferred sessions back to the mobile device.

Furthermore, the corresponding party should not be aware of a session transfer going on; nor should the corresponding party's device require a specific update of the SIP client software. All functions on the originating-party side should be performed with minimal user input.

The following example scenarios further illustrate the functionality.

- *Profile mobility* (1). An airport public telephone booth contains a phone and a separate ID card reader. Tony, a traveller, approaches the booth and inserts his personal ID card, which contains his SIP address and authentication information. The token reader registers him for the booth with his SIP address as a contact. Automatically, the phone is updated with Tony's profile. Tony now has his home server set as the outbound proxy and his address book and other personal items set on the phone. When Tony inserts the ID card a second time, the reader initiates removal of the configuration from the public phone in the booth.

- *Profile mobility* (2). While Charlie is using the PC soft phone in the conference room, he adds a number of address book entries. Each change updates his roaming profile, which causes a notification to be sent to the PDA that he is carrying, synchronizing the devices without his request. He leaves the room with his PDA, forgetting to take any explicit action to remove his configuration from the PC soft phone, such as swiping his identification card at the door. The soft phone is subscribed to Bob's presence and therefore tracks his location. When his PDA updates his location in another room after receiving the coordinates from a Bluetooth beacon, the soft phone recognizes that he is no longer in the room and removes his data so that another user may not view it. As a result of the synchronization, his changes to the address book are also available on his PC.

- *Session mobility*. Alice is on a business trip to visit a subsidiary of her company in Paris. While she is still on her way, she receives a video conference call from her partner, which she accepts on her mobile phone. Arriving at the office, she immediately is directed to a meeting room equipped with diverse communication facilities, which are discovered by her mobile phone when she enters the room. In order to get a better resolution she transfers the video stream to a local display. As this device has no audio capability, she transfers the audio part of her session to an IP phone on the table, using its loudspeaker and microphone for hands-free discussion.

With these requirements, our architecture, shown in Figure 3-22, consists of three basic elements: signalling (solid line) for profile and session transfer between devices (D1, D2), servers (Proxy) and device managers (CD, DM), the multi-devices system (DM) being used

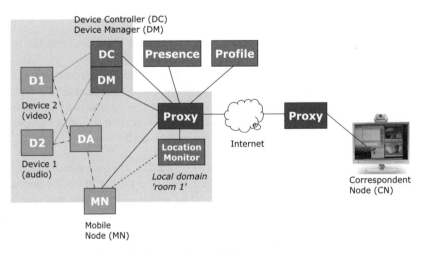

Figure 3-22.   Session mobility system overview

to shield a split transfer from the corresponding party, and location-based device discovery (Directory Agent (DA), shown by the dashed line).

Several solutions have been proposed for service mobility as it is a widely discussed example of ubiquitous services in many research projects. However, most existing solutions are either based on proprietary protocols[26] or on mechanisms requiring new software implementations in endpoints; alternatively, they introduce new elements such as media proxies that have to be inserted in the media stream prior to session transfer.[30] Other solutions concentrate on specific service types such as media streaming.[27] Looking at the protocol that is used for IP-based service platforms in mobile communications systems such as IMS, which is the Session Initiation Protocol, we can see that it already includes basic mechanisms for signalling service and profile transfers.

The solutions used as an example in the following sections are based on standard SIP with standardized packages such as the SIP event framework. Non-standard extensions to the SIP protocol are not required. Furthermore, the called party is not aware of the transfer except for the change of media stream endpoints, which is handled automatically by SIP.

3.3.3 Realizing Session Mobility

Session mobility describes the transfer of active sessions between devices, including the possibility of splitting the session between devices. In a mobile environment, we refer to the transfer of ongoing sessions between a mobile device (Mobile Node, MN) and a Correspondent Node (CN) from the MN to devices acquired from the local environment and also to returning them back to the MN.

Two modes of session transfer can be distinguished: Mobile Node Control and Session Handoff. In Mobile Node Control (MNC), the mobile device always has direct control of the session signalling, even though the media are transmitted between the local device and the CN. In the Session Handoff (SH) mode, the mobile device relinquishes control of the ongoing session after transferring the session to a local device. A new session between the CN and a local device is created and this new session will replace the ongoing session between the mobile device and the CN.

Furthermore, each of the transfer modes also has two options, either to transfer the whole session or to split a session while transferring to separate devices. Whole Session Transfer is a simple case of session transfer, where an entire multimedia session is transferred from one device to another single device. For example, a voice call could be transferred to an IP phone or a voice and video call could be transferred to a video phone. If a local device supports a type of media that is not available on the user's mobile device, he could move the session and add the media type. This would be done if the mobile user walked into his office and decided to take advantage of the video phone in the room to enhance his currently audio-only call. A transfer could also reduce the number of media in the session if the user, for instance, decided to transfer his audio and video session to an audio-only IP phone.

In Split Session Transfer, a multimedia session is split across multiple devices. This may include the Mobile Node if, for instance, the user transfers one media type to another device, while keeping the other local, or adds another media type to the session by inviting another device. Alternatively, all of the session's media may be transferred to other devices. The session could be further split by dividing a single medium among multiple devices. For instance, the user of a mobile phone or PDA who is in a video session may be satisfied with transmitting his own video with the device's camera but may find the display too small for properly viewing the other call participant. He will, therefore, transfer the video output to a projector in the room while keeping the input on his mobile device, with both operating in simplex mode.

As described earlier, in the architecture presented here, the SIP suite is used as a signalling protocol to transfer ongoing sessions from the Mobile Node (MN) to the local devices and also to return them to the MN.[31,32]

3.3.3.1 Session Transfer in the Mobile Node Control Mode

In the example described here, for Mobile Node Control we use SIP Third Party Call Control (3PCC).[33] Here, the Mobile Node acts as the controller to establish a separate session with the local device and updates its session with the CN in order to establish a media flow between those two endpoints. The MN sends an INVITE to the local device, requesting it to establish a new session. The local device responds with its media parameters in the SDP body, and these are then used to re-invite the CN. Once the CN and the local device have each other's media parameters, they create a direct media session between them. In order to split the session across multiple devices, the MN establishes a new session with each local device through a separate INVITE request and updates the existing session with the CN, based on the media parameters it receives in the SDP bodies of their responses.

3.3.3.2 Session Transfer in the Session Handoff Mode

For session handoff the architecture presented here uses the SIP REFER method[34] to request that a new session is created between the local device and the CN to replace the existing session. The MN sends a REFER to a 'referee', which can be either a local device or the CN, requesting it to contact a specified target URI. For the architecture presented here, the REFER request was chosen to be sent to the local device. Upon receiving the request, the local device sends an INVITE request to the CN to initiate a new session. The 'Referred-By' header is used to identify the MN as the requestor of the new session. The 'Replaces' header is used to request that the new session replaces the existing session between the CN and the MN.

There are three ways of providing Splitting Session Transfer by using the REFER method, as described below. In each case, we assume that a session between two parties, namely the 'Mobile Node (MN)' and the 'Correspondent Node (CN)', has been established and the MN discovers the local devices (one for audio and another for video) available in his area and would like to make use of them to continue his ongoing video telephony session.

In the first alternative, the MN sends a REFER message to the CN asking the CN to send an INVITE message to local devices to invite them into a session. After those local devices have accepted the call from the CN, the CN will inform the MN about this new session, but still on behalf of the MN by using the NOTIFY method. Finally, the session between the MN and the CN will be terminated. A disadvantage of this method of transferring the session is that it appears to the CN as two new calls, rather than just a device transfer at the other end.

The second possibility would be for the MN to send a REFER message to each of local devices to ask them to send an INVITE to the CN, including several headers (for example a Referred-By header and a Replaces header), the necessary information about the current session (dialog, call-id, tags, etc.) and the MN's authentication body. This authentication is used to authorize the local devices to replace the current session and the CN will accept the INVITE sent automatically by local devices. This way of session transfer is more seamless than the first one, since the CN merely needs to update the new media connection without making any new calls. Nevertheless, the CN still has to terminate separately the call to those local devices, since the call transfer still appears to the CN as two different calls.

The third option introduces a Device Manager (DM) as a separate functionality, to which the REFER is sent. The DM is used to create a 'multi-devices system', which is a single system that connects many devices. This DM has a SIP Back-to-Back User Agent (B2BUA), which has the functions of both a SIP UA and a SIP proxy server. In this case, the MN only sends a REFER message to the DM, and the DM sends an INVITE to the audio and video devices. These local devices will response with the 200 OK message containing an SDP body message specifying the preferred media connection (for example the host address) and media parameters (the media name and transport port). The DM then sends an INVITE to the CN including the new media information for the audio device and the video device. As in the previous case, this INVITE message also contains the necessary information and headers to replace the ongoing session between the MN and the CN with the new one. Figure 3-23 illustrates this alternative.

If these three ways of providing splitting session transfer are compared, using the DM seems to be the most seamless and simple way from the CN's side. In particular, the CN does not have to terminate separately the call for each local device, since there is only one control session between the DM and the CN.

3.3.4 Discovery of Ubiquitous Devices

As organizational domains may be very large and devices may be frequently added and removed, the dynamic and decentralized maintenance of accurate device location information is therefore of increasing significance.

In the architecture presented here, the Service Location Protocol (SLP)[35] is used for dynamically discovering contact information for all the SIP-enabled devices available in a user environment. In general, any other service discovery protocol could be used. The following illustrates the problems and possible solutions in a SLP example. To achieve this, the 'sip' service type for all SIP endpoints in the network must be defined and each service available on each endpoint is identified by using the standard SIP URL, such as sip:audio@ dev1.example.com. The service attributes specifying device characteristics (for example vendor, supported media or codec) and location parameters (such as room number) are described in the SLP Service Template.

In a small network, the service information is normally stored in the SLP Service Agent (SA), which could be either co-located on the same host, such as a PC, as that on which the SIP UA is running or on a separate host, as in the case of a hardware device like an IP phone. However, for a large organization, SLP has an option to use the Directory Agents (DA); this option uses central data memory and provides scalability and robustness to the networks. In

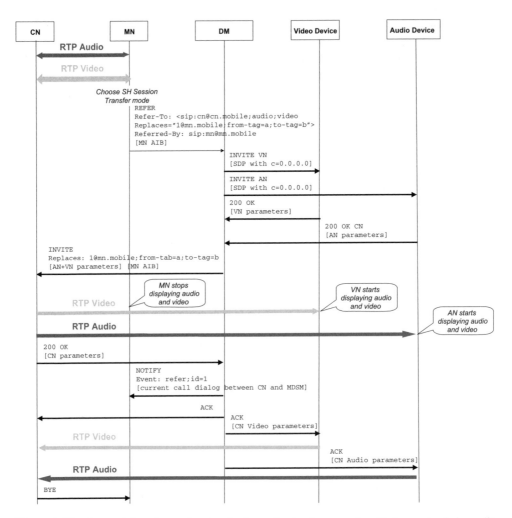

Figure 3-23. Protocol flow for session transfer in session handover mode with the session being split beween two devices

this case, SLP SA will send services information to the DA by sending a service registration (SrvReg) message to the DA. After the registration is processed, the SLP User Agent (UA) on the DC will be updated about the availability of all devices whose locations are in the area served by the DC.

To be able to discover the service provided by the multi-devices system, the Device Manager (DM) must first find all available services in its area by sending the SrvRqst to the DA. Then the DM creates a local 'username' identifying the new service (for example sip: audio_video@dc.example.com) and sends a SrvReg message to register this service, which lists the location and media attributes of the system.

When a mobile user carrying a mobile device comes into a room, the SLP UA on his device must first discover the DA either by sending a multicast Service Request (SrvRqst) using the Directory-Agent service or by trying to obtain SLP options from the Dynamic Host Configuration Protocol (DHCP) server. Once the mobile device knows the location of the DA, the SLP UA on the mobile device sends a request (SrvRqst) for the 'sip' service to the DA and waits for the reply (SrvRply), which contains lists of all SIP services including service URLs and service attributes. To filter out non-local devices in the response, the location attributes of the mobile device should be sent along with the SrvRqst message.

3.4 Summary

The service platform architecture is going to be of a greater importance than ever before in future mobile communications systems. Not only will network operators have to focus on services and applications in order to differentiate them from competitors, but platforms will be opened to third-party providers to enrich the service portfolio. Sophisticated business models, such as that shown in the i-mode, provide a win–win situation for all sides, including the user, who can choose from a richer service offering, which may even be personalized. In fact, personalization is regarded as a major driving feature for service platforms to foster customer retention. Looking at new service platform capabilities and paradigms, we will immediately see that the area of ubiquitous communications is a new field where diverse services are provided, mostly by non-traditional operators. In order to provide a seamless service experience, heterogeneous ubiquitous environments have to be considered for inclusion in a future service platform.

In this chapter, we have explained the whole area of service architectures and, in particular, mobile service platforms. The different platform architectures illustrate concepts and paradigms for service provisioning and also underline the increasing importance of service platforms in general. Based on this discussion of the state of the art, we have outlined the capabilities envisaged for future mobile service platforms. In particular, personalization and context awareness must be established for user-centric service provisioning. Ubiquitous services extend the traditional cellular-network-based mobile service provisioning towards ubiquitous communications environments. We have given a detailed system example for a concrete realization of session mobility in an IP-based service platform using SIP.

4

Extension to Ubiquitous: Mobile Peer-To-Peer

Zoran Despotovic and Wolfgang Kellerer

P2P computing is emerging as an important paradigm that can have a disruptive impact on the service platform. It can enable a whole new range of applications and bring the service platform to areas which it was hard to imagine just a few years ago.

Over recent years peer-to-peer (P2P) computing has been receiving enormous attention, both in academia and among major industrial players. It is believed to be a disruptive technology that may enable a range of new applications and impact the future of information technology. Today's P2P systems such as Gnutella, Edonkey and BitTorrent are all good examples of a P2P success story: rather simple software enables Internet users to exchange files. The popularity of these applications has gone so far that many empirical studies (see Sen and Wang,[1] for example) report that the P2P traffic represents an overwhelming proportion of the overall traffic in the Internet.

Although it is file sharing that has driven the development of P2P systems, the promises and potential of P2P technology go far beyond this simple application. In recent years we have been witnessing the development and wider adoption of a number of other applications. Examples include P2P web caching,[2] P2P information retrieval[3] and decentralized storage systems.[4] It also turns out that the P2P paradigm provides elegant and efficient application-level equivalents of a number of communication primitives that are normally implemented in the lower layers of the communication protocol stack. Application-layer multicast and anycast are typical examples.[5,6,7] Stoica *e al.*[5] go even further and provide support for mobility with the proposed P2P-based Internet indirection infrastructure. The lessons learned from successful P2P design have recently led to a definition of new routing protocols for mobile ad hoc networks strongly based on Distributed Hash Tables (DHT) principles.[8] In a word, after a few years in which the outlook was unclear as a result of copyright issues associated

Towards 4G Technologies. Edited by Hendrik Berndt.
© 2008 John Wiley & Sons, Ltd.

with file sharing applications, it seems that P2P technology has won the battle and is now promising to have an important impact on information and communication technology.

P2P systems offer an alternative to traditional client–server systems for many application domains. This is particularly true for Internet-scale distributed environments. In such environments resources are concentrated on a small number of servers, which have to serve an extremely high number of client requests. As a result, they must apply sophisticated load-balancing and fault-tolerance algorithms so as to provide continuous and reliable access. In addition, their network bandwidth must be increased from time to time as the client base grows steadily. In contrast, in P2P systems every node (peer) in the system acts as both client and server, contributing in part to the successful operation of the entire system. Put another way, every node *pays for* its participation in the exchange community by providing access to its computing resources. Thus, the P2P approach circumvents many problems of client–server systems. The most important among these are scalability, fault tolerance and maintenance costs through self-organization.

We begin the chapter by defining P2P systems. The exposition is based on Aberer *et al.*[9] and Despotovic.[10] Although it has a rather technical flavour, motivated by our wish to establish a good base for later discussion, we believe that it is at the same time simple and easy to understand. Then, we discuss unstructured P2P systems and structured P2P systems (Distributed Hash Tables) as the two main classes. Considerably more space is devoted to DHTs because they outperform the unstructured systems. In the second part of the chapter ,we focus on the specifics of applying P2P technology to mobile environments. We first give the motivation for this step by describing a number of potential mobile P2P applications. Then we outline the main challenges that have to be met in order to apply P2P successfully to mobile settings. Finally, we describe existing solutions to some of these challenges and also provide hints on how to solve those for which we could not find ready solutions.

4.1 P2P Systems Definition and Main Classes

In short, a P2P system is an application-layer resource-lookup system in which participating peers establish logical connections with one another in order to enable an efficient location of the considered resources. Before we go into more detail, let us explain the motivation for P2P research by saying why we might need such application-layer lookup systems. The answer is rather simple: because there is no appropriate support for the resource location task in the lower layers of the protocol stack. (Keep in mind that the word 'resources' is used here to denote application-level concepts.) Files in an unavoidable file-sharing application are a good example. Data items representing observations of sensors connected to a set of computers are another example. Imagine now that the considered resources are distributed among a large number of computers across the Internet. Clearly, the IP protocol can route among different computers, but it is unaware what data resides at which machines. On the other hand, it turns out that large-scale decentralized applications based on producing and disseminating enormous amounts of data are of great interest for a large user base; hence the need for application-layer resource-lookup systems as the main building block for such applications.

As we have said above, the central problem for P2P systems is resource location without centralized control. To establish the notation, let us assume that we deal with a set of resources that are distributed across a set of peers P (we will use the symbol $p \in P$ to denote the peer

p's physical address too; the meaning will be clear from the context). Let R denote the resource set. Assume also that the individual resources are identified using application-specific identifiers from a key space K. We will assume that the key space is equipped with a distance function, $d: K \times K \to \mathbb{R}$, which describes how far apart are any two specific keys. As mentioned above, the resources can be anything, files shared in the system, various data items or even computing cycles. The keys can be thought of as, for example, positive integers or strings representing the names of individual resources.

The problem to solve is then the following: any peer participating in the system should be able to locate as efficiently as possible a peer with physical address $p \in P$ holding a specific resource $r \in R$ that is identified by a key $k \in K$. In order to perform this task we assume that the peers maintain logical, application-level connections to a small number of carefully selected other peers and can forward resource requests to them. Put another way, the peers construct an application-layer *overlay network*. This overlay network should allow addressing of resources by their application-specific keys, instead of using application-independent physical peer addresses. The problem is how to realize such an overlay network efficiently, i.e. how to realize basic operations of overlay network maintenance and routing (such as inserting new files and searching for files in the case of a file-sharing application) with low consumption of physical resources, for example the network data rate.

We briefly summarize the above reasoning by outlining the main components of a P2P system:

A P2P system is a tuple $(P, R, K, G, \text{RA}, \text{MA})$, where:

- P is a set of peers;
- R is a resource set to be distributed among the peers P;
- K is a set of keys used to identify resources;
- G is a directed graph (P, V) representing the overlay network;
- RA and MA are algorithms operating on the graph G and enabling resource lookup (RA) and network maintenance (MA).

Assuming that we have constructed an overlay network G, the routing algorithm RA can be thought of as a mapping *forward*: $P \times K \to 2^P$ with the interpretation that *forward*(p, k) is a subset of the peers to be contacted via outgoing edges of p in G when p is asked to locate the resource identified by key k. We emphasize that the outdegree of the nodes in the graph G should not be too high, since the purpose of the maintenance algorithm MA is to keep the outgoing links consistent in the presence of peer joins, leaves and failures. Too high an outdegree would require too costly maintenance.

Based on this abstract view, two important classes of P2P systems can be distinguished.

- *Unstructured P2P systems*. The distinctive property of the unstructured P2P systems is that there is no correlation between the distribution of resources over the peers and the structure of the overlay network G; the links among peers are established randomly, i.e. independently of the resources the peers are responsible for. Therefore, the resource lookups are performed in an exhaustive fashion, for example by using broadcast or flooding of the search requests. Examples are Gnutella[11] and Lv *et al.*[12]
- *Structured P2P systems* (also known as Distributed Hash Tables, DHTs for short): Unlike unstructured systems, structured P2P systems are characterized by a dependence of the set

of resources that are any peer's responsibility and the links that the peer maintains toward other peers. Normally, every peer is responsible for a group of 'adjacent' resources and links are established to 'nearby' peers and to a small set of 'remote' ones. (Note that all quoted words are closely related to the mentioned distance function.) In this way, lookups can be performed in a greedy fashion, by forwarding the request to as close to the destination as possible. Examples of structured systems include Chord,[13] CAN,[14] [Ratnasamy01], Pastry,[15] Viceroy,[16] Symphony[17] (based on the small world graphs theory of Kleinberg[18]) Kademlia[19] and PGrid.[20]

Figure 4-1 gives an overview over the different P2P concepts classified in unstructured and structured P2P. Note that some authors mention hierarchical systems (see sGarcés-Erice et al.[21] and Mizrak et al.[22] for examples) as a third class. We do not see hierarchical systems as a separate class, mainly because they do not introduce any new dimension in the system design; they rather combine existing solutions from the two mentioned classes to solve specific problems. We show in Section 4.3.3 what advantages these concepts have in heterogeneous system environments.

Note also that we do not explicitly include centralized systems (such as Napster) either. Such systems rely on a centralized index for the resource lookup phase. The resource exchange is then performed in a truly P2P manner.

The next two sections give details of the two classes mentioned. In particular, we provide more information on the DHTs because of their apparent advantages over unstructured P2P networks. (These advantages will become clear as we proceed, if they are not already clear.) To provide a precise comparison of unstructured networks and DHTs, as well as various DHT

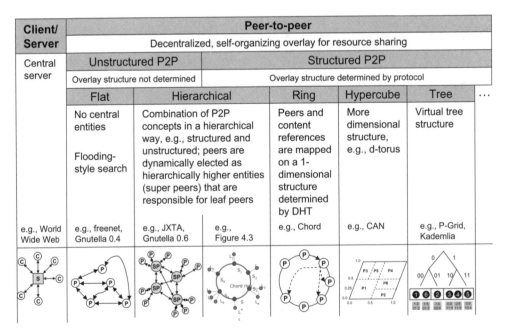

Client/ Server	Peer-to-peer					
	Decentralized, self-organizing overlay for resource sharing					
Central server	Unstructured P2P		Structured P2P			
	Overlay structure not determined		Overlay structure determined by protocol			
	Flat	Hierarchical		Ring	Hypercube	Tree ...
	No central entities Flooding-style search	Combination of P2P concepts in a hierarchical way, e.g., structured and unstructured; peers are dynamically elected as hierarchically higher entities (super peers) that are responsible for leaf peers		Peers and content references are mapped on a 1-dimensional structure determined by DHT	More dimensional structure, e.g., d-torus	Virtual tree structure
e.g., World Wide Web	e.g., freenet, Gnutella 0.4	e.g., JXTA, Gnutella 0.6	e.g., Figure 4.3	e.g., Chord	e.g., CAN	e.g., P-Grid, Kademlia

Figure 4-1. Peer-to-peer concepts

proposals, we will evaluate the proposed solutions with respect to how they answer the following set of questions:

- What is the structure of the overlay network and how are lookups performed?
- What is the routing complexity, i.e. how efficient are the lookups?
- What is the network maintenance overhead and how efficiently can peers join and leave the network?

For a more detailed analysis of these questions, as well as analysis of other important issues, such as load balancing, security and proximity-based routing, just to name a few, see a number of overview papers that have appeared recently.[23,24,25]

4.1.1 Unstructured P2P Networks

We said that the distinctive property of unstructured P2P systems, compared to structured P2P systems, is that peers do not commit or restrict themselves to managing any specific set of resources, so that the distribution of resources over the peers and the structure of the overlay network G are independent. Thus any peer can hold any resource from R. This is just another way of saying that the peers choose their resources independently of one another and connect to a number of other peers at random, without considering what resources their neighbours hold. As a result, in order to obtain access to resources managed by other peers, peers use the graph G to implement a request-forwarding scheme that has to be designed such that it will eventually reach any peer in the P2P network.

Gnutella[11] is a typical representative of unstructured P2P networks. In Gnutella each peer is connected to a small number of neighbours, a typical number being four. Resource lookups are performed by flooding the requests through a large part of the overlay network, starting from the requestor and forwarding requests recursively to all known neighbours, neighbours of the neighbours and so on. In order to control the number of messages, the flood has an upper limit, the so-called time-to-live (TTL), a typical value being seven. In addition, request messages carry identifiers so that requests returning along a cycle to a peer that has already seen the message can be dropped. Because typical values of the TTL are set so that almost the entire overlay network is explored, we can see that the number of messages generated by the broadcast is linearly proportional to the size of the network (the size here denotes the number of edges, not nodes), meaning that almost every pair of connected nodes will exchange a message in order to handle a single search request. Clearly, this is an extremely high communication overhead and we can immediately say that this is the most severe drawback of the unstructured overlay networks. However, it is important to understand that the search latency (which is perhaps most important for user perception) depends on the diameter of the overlay network. As it turns out, for random graphs and small world graphs, one of whose representatives is Gnutella, the diameter is logarithmic in the number of nodes. This means that Gnutella provides a sufficiently small search latency.

A similar mechanism is used in Gnutella for constructing and maintaining the network. Peers joining the network flood the network with a discovery message. From the responding peers the joining peer then selects its local neighbours. If peers fail to operate they are dynamically replaced from a list of known peers in the network. It has been shown that this mechanism leads approximately to a power-law distribution of incoming links, as peers tend to attach preferentially to stable peers. Similar behaviour has been discovered in many self-organizing networks, including the Web.

We have said that the flooding is fairly expensive: a single search request generates load for a large portion of the peers. This is why there have been a number of attempts to improve the lookup costs, expressed in terms of the number of generated messages, by improving the lookup request forwarding scheme. A group of these approaches is based on the so-called random walk: the requests are not forwarded to all the neighbours, but only to a randomly selected one, which gives rise to a depth-first search strategy. Another group of approaches uses results from percolation theory to estimate precisely the number of links that need to be traversed in order to reach the whole network. Even though they outperform Gnutella, the search efficiency of these approaches remains unacceptably low. Essentially, just like Gnutella, they lack coordination among the peers, i.e. knowledge of which peer is managing which resource. It is exactly this problem that makes search in unstructured P2P networks inherently expensive in terms of communication overhead.

However, unstructured networks also have a number of desirable properties. Most notably, updating a resource and inserting a new one are literally trivial operations and have no cost at all. When, for example, a new resource is inserted into the system, the owner of the resource does not need to inform any other peers about it. A similar statement is true when it comes to the maintenance of the network structure. When a peer leaves the network it does not need to inform others. These are the good sides of there being no dependencies among peers and between peers and resources and thus these operations can be performed autonomously.

Another good property of unstructured networks, which we will fully appreciate after we have seen that it takes some effort to realize it in the DHTs, is the ease of handling complex requests, such as complex queries. For example, if the resources are annotated with a set of attributes, multi-attribute queries are natively supported. This is not the case with the DHTs.

4.1.2 Structured P2P Networks – Distributed Hash Tables (DHTs)

Unlike unstructured systems, there is coordination among the peers in a structured P2P system. In this section we describe how this coordination is established and what benefits it brings, but also what price has to be paid for having these benefits.

In a DHT every peer commits to managing a specific subset of resources. In general, the distribution of the resources depends on the set of all peers present in the system at a given time instant. It is carried out in the following way. Each peer first becomes associated with a key taken from the key space K (recall that K contains the resource identifiers). We can think of this step as peers picking keys at random. Normally, the size of the key space is much bigger than the number of peers so that no two distinct peers are expected to select the same key. At the same time, each key is associated with a partition of the key space such that the peer holding the key becomes responsible for managing all resources identified by keys from the associated partition. Typically the key partitions are created by selecting all the keys that are in some appropriate way closer to the key of the peer in question than to those of other peers currently in the system. This is why the key space K needs to be equipped with a distance function d. All this reasoning is actually nothing else than introducing the following two functions: a function $key\colon P \to K$ that associates peers with keys and, given $key(P)$, a function $partition\colon K \to 2^K$ associating peers with partitions of K.

In order to forward resource requests efficiently, peers form a routing network G by taking into account the knowledge about the association of peers with key partitions. This network can be defined as an additional mapping, *neighbours*: $K \rightarrow 2^P$, which associates peers with their neighbours in graph G. The distance function d plays an important role in constructing the network. Typically, peers maintain short-range links to peers with neighbouring keys and, in addition, a small number of selected more distant peers, where the probability of having a long-range link to a peer decreases with the distance to the peer's key. The standard approach that has been used in most structured P2P networks is to choose long-range links with exponentially decreasing probability depending on the distance. All outgoing links of a peer make up what is usually called its routing table in the P2P literature. Using routing tables, peers then forward resource requests in a directed manner to the closest peers that they know from their routing table, i.e. trying to greedily reduce the distance to the key that is being looked up.

Let us explain in an example how all this works. Chord is one of the most popular and simplest DHTs.[13] We can think of the keys just as ordinary integers (as opposed to strings or points in a Euclidean space, for example) that are ordered around a circle, as shown in Figure 4-2. The ticked, topmost point of the circle corresponds to the key space size and all arithmetic is modulo this number. The metric d is defined simply as the clockwise distance between two points. The key partitions are created so that any peer is responsible for the arc starting at that peer and going counter-clockwise to the next available peer (but excluding it). Thus, the peer P14 from Figure 4-2 is responsible for keys 14, 13, ..., 9. Figure 4-2 also shows the routing table of peer P08. The peer maintains links toward peers following the keys that are powers of two away from P08. Thus, a link is maintained to the immediate successor on the circle because this successor must be responsible for the key $8 + 2^1 = 9$. The same peer turns out to be responsible for the keys that are 2 (i.e. 2^1) and 4 (2^2) away. The longest-range link is established toward a peer following the key that is half the key space away. This is why P08 has a link to peer P42, to whose partition the key 40 ($8 + 2^5$) belongs. If peers are linked in this way, then routing becomes extremely simple: the request is forwarded by following the longest possible link without exceeding the key to which the routing is aimed. The peer that would exceed the key just selects its immediate successor as the target peer. To route to a key, say 54, peer P08 uses the longest-range link to reach the peer P42 and this peer then forwards the request to peer P51 (going farther would exceed the key). Finally, P51 just forwards the request to its successor P56.

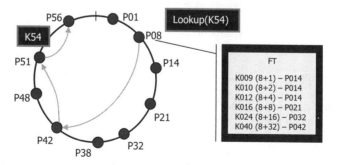

Figure 4-2. Chord DHT (Reproduced from reference [13] by kind permission of the authors and ACM Publications, New York.)

Chord is an example of systems that use a deterministic neighbours function. With deterministic functions, the structure of the overlay network is known in advance, given the sets of participating peers. At the opposite end are randomized systems in which this is not the case.

It should be obvious that routing in Chord is logarithmic in the size of the network (i.e. the total number of peers) because every hop halves the remaining distance to the destination. At the same time, routing tables are also logarithmic in size. Needless to say, this is acceptable for most of the settings. For large file-sharing networks with millions of nodes, this means that any file can be found with only a few hops, while each peer maintains only a few routing entries. These two properties are common for most of the proposed systems. However, it is worth mentioning that recently there has also been some work that achieve constant-outdegree graph topologies and consequently constant-sized routing tables while retaining logarithmic routing. Two well known examples are Viceroy,[16] using so-called Butterfly topology, and Koorde,[26] based on de Bruijn graphs.

Generally, the specifics of any DHT solution can be neatly presented by describing its choice of the key space, distance function and linking strategy. These have been the subject of intensive research over recent years, which has resulted in numerous designs of structured-overlay networks. We summarize in Table 4-1 the main properties of some representative

Table 4-1. Main properties of various DHTs

P2P System	Comparison criteria				
	Geometry*	Distance	Parameters	Hops to locate data	Routing state
Chord	Ring	Clockwise integer difference	N = #peers	$O(\log N)$	$O(N \log)$
CAN	Hypercube	Euclidian distance in a d-dimensional space	N = #peers d = #dims	$O(d \cdot N^{1/d})$	$O(d)$
Pastry	Hybrid (tree–ring)	Height of the smallest subtree containing both or clock-wise integer difference	N = #peers b = id base	$\log_{2b} N$	$b \cdot \log_{2b} N + b$
Kademlia	Tree	Integer value of the bitwise XOR	N = #peers	$O(\log N)$	$O(\log N)$
P-Grid	Tree	Height of the smallest subtree containing both	N = #peers	$O(\log N)$	$O(\log N)$
Viceroy	Butterfly	Clockwise integer difference	N = #peers	$O(\log N)$	$O(1)$
Symphony	Kleinberg small world on a ring	Clockwise integer difference	N = #peers k = outdegree	$O((\log^2 N)/d)$	$O(d)$
Koorde	de Bruijn	Clockwise integer difference	N = #peers	$O(\log N)$	$O(1)$

* As Gummadi points out,[23] the notion of geometry may be lacking a precise definition. However, we rely on the reader's intuition and hence we use it here too.

solutions that have been proposed. Note also that some systems are designed as to allow tradeoffs between routing complexity and routing state. For example, Koorde can also route in O(log N/log log N) hops with the state O(log N).

4.2 Some Problems with DHTs

In this section we will elaborate on some seemingly simple problems that require non-trivial DHT solutions, outlining in a sense the price we need to pay for having highly efficient DHT routing. First, as a result of the dependencies introduced among peers, network maintenance is required to keep routing tables consistent when peers join or leave the system. The main reason for keeping the sizes of routing tables small is actually not the required storage over-head, but rather the maintenance cost incurred by having large routing tables. Second, unlike unstructured P2P networks, changes to the resources themselves, i.e. insertions, updates and deletions, are not localized to the peers maintaining the resources, but affect other peers as well. Third, resource requests are constrained to simple key-based lookups. Even slight generalizations, such as requesting ranges of keys, require non-trivial algorithms for implementation.

4.2.1 Maintenance Overhead

If we were to make a judgement based on the properties we have discussed so far, DHTs would appear to be a perfect solution for large-scale P2P systems because they can route efficiently while maintaining small per node state. However, the good properties related to the efficiency of routing do not come for free. To construct and maintain a structured P2P network, peers have to deal in particular with the problem of node joins, leaves and failures.

Joining usually coincides with a routing operation and as such has logarithmic complexity. To join the system, peers usually have their own key routed via one peer they know in advance. In this way they learn about their position in the DHT and also about peers whose routing tables they can use to construct their own. As a final step, a joining peer receives references to a set of resources for which it becomes responsible. These references are obtained from the set of peers closest to the joining peer's key. The operation of leaving the network implies similar steps, the main goal of which is to maintain the network consistency. For instance, the nodes pointing to a leaving peer should be informed about its departure so that they can update their routing tables. Thus, joining and leaving the network are clearly operations that generate some overhead. It is normally not too high but it is not negligible either.

Another source of overhead is potential failures of DHT nodes. These are different from node departures because they do not involve steps to ensure the network consistency. The problem is that inconsistencies may lead to serious degradation of DHT performance. In the worst case the network may become partitioned so that no routing is possible between groups of nodes. This is why additional algorithms have to be run at the nodes to maintain the network consistency and re-establish it when it is lost. They normally involve probes by each node of all the entries in their routing tables, followed by attempts to replace every failed entry with an existing node. If the failure rate is high, then these steps have to be executed frequently, which in turn can have the undesirable effect of amplifying the failure rate itself. Thus, the failure detection and recovery algorithms need to be carefully designed. See Rhea et al.[27] for a detailed analysis of this problem.

4.2.2 Complex Queries

DHTs natively support only exact-match lookups: in order to search for a resource, one has to know the key associated with the resource. However, most of the time the exact keys are not known to the query issuers so that key-based search is not enough. This seriously limits the DHTs' applicability in practice. Typically, every resource is annotated with a number of attribute–value pairs and users want to be able to execute more complex queries, collecting all objects with specific values of a set of attributes. For example, in a file-sharing P2P network many search requests would be comprised of a fraction of the file name, a file type (such as an audio or movie file) and the performer's name (in the case of audio files). DHTs are not designed to support this type of query.[28]

So-called range queries are not supported either. They return all objects with keys or values of any other attribute falling within a specific range of values. The exact problem is that the keys of the resources are derived independently of the values that attributes of those resources take. For example, the keys can be obtained as SHA-1 hashes of (a part of) the binary representation of the resources. If this is done, semantically close objects, i.e. those having just small differences in the values of one or more attributes, have totally different hash values and are thus spread over the network in an uncoordinated manner. The actual reason for using such hash functions lies in the designer's wish to provide a solution for load balancing: the peers should share the load as evenly as possible, i.e. they should be responsible for approximately same number of keys. On the other hand, the order among the values of an attribute can be preserved if an order-preserving hash function is used. But this can hurt load balancing because a skewed distribution of the attribute values leads to skewed key partitions.

Thus it turns out that the two problems, load balancing and support for range queries, are incompatible; solving one makes the other worse. An example of a system giving a higher priority to range-queries support is P-Grid,[20] which uses an order-preserving hash function and a load-balancing strategy that is different from those of other systems. The algorithm implementing range queries in this system is described by Datta et al.[29] Triantafillou and Pitoura[30] follow similar ideas for implementing range queries but also offer the argument that good load balancing can be achieved by employing replication (which is needed anyway in practical systems for higher fault tolerance). Resources are normally annotated with multiple attributes and load balancing can be achieved by using different hash functions for different attributes.

Most systems, however, opt to solve load balancing as described above and provide additional support for range queries. Ramabhadran et al.[31] solve the problem of range queries by superimposing a 'prefix hash tree' (PHT) on raw (unhashed) values of a considered attribute on top of an existing DHT, trying thus to recreate the ordering lost by hashing the attribute values. Compared to P-Grid, in which range queries are natively supported, PHT is less efficient. The primary reason is that, given a value to lookup, many prefixes of this value have to be searched for because PHT does not maintain such information. On the assumption of binary search on different prefixes and the size of the attribute space being 2^D, the number of PHT operations becomes $O(\log D \cdot \log N)$. However, it is important to note that PHT does not make any assumptions about the underlying DHT (it can be implemented on top of any DHT) and it does illustrate that range queries are possible in DHTs. In our opinion, these are its main advantages.

What solutions are appropriate for more complex types of queries such as multi-attribute queries, which return all resources for which the values of a group of attributes satisfy a specified predicate? (We still assume a simple data model in which every resource is associated with a number of (*attribute, value*) pairs.) The most common solution is to index each of the involved attributes and then run the query for each of the attributes appearing in the query. The final query processing, i.e. filtering of the returned results, can be done locally, at the querying peer. Even though distributed query processing is an area investigated extensively by the database research community, there are few applications of its results that optimize this P2P query-processing strategy. Furthermore, the task of selecting the right attributes to index is important. Those never or rarely queried should not be indexed; see Klemm et al.[32] for more details.

A generalization of the ideas we have just discussed applies when the resources available in the system have a more complex structure, though the problem becomes somewhat more complicated in this case. Without loss of generality, we assume that every resource is associated with an XML description. These descriptions can be arbitrarily complex and their schemas need not be the same. A standard approach in this case is to generate an auxiliary index (metadata) describing queries that can return some result, i.e. providing a mapping between 'general' queries that can be executed by users and objects to be returned as their results. This auxiliary index itself is stored in the underlying DHT as an additional data set, in which queries are considered as keys. The query process then proceeds in the following two phases. Given a query, the auxiliary metadata is queried first to discover the keys of the objects that will be present in the result set. Then the DHT is queried for these keys to find the locations of the objects themselves. This may actually involve more than two phases, so that we go from more general to more specific queries until we eventually reach the exact match queries. Garcés-Erice et al.[33] and Skobeltsyn et al.[34] present two examples using this approach. Though this simple idea provides a solution to the problem of executing complex queries, it is not a complete solution. The most severe problem is that, even for simple forms of data, it is very hard to predict what queries can be posed on the data. Theoretically, the number of possible queries can grow extremely large and maintaining the index including all of them would be a burden, if not impossible. This is why it becomes important to be able to predict the query distribution, i.e. what queries will be executed by the users along with their frequencies. Once an appropriate solution to this problem is found, the challenge moves to reducing the costs incurred in the index maintenance itself and finding the best trade-off between the benefits the index brings and the costs it incurs.

4.3 Mobile Peer-to-Peer

We now move our attention to the specifics of applying to mobile environments the P2P concepts we have discussed so far. We will briefly outline what benefits the P2P technology can bring, not only to mobile users and operators but also to potential service providers. We will also mention a range of useful applications that might be implemented in a P2P style. In the second part of this section we first discuss the challenges that the P2P implementation of such applications raises and then outline possible solutions and ideas to meet those challenges.

Before we proceed, we provide a simple, but quite useful, characterization of mobile P2P systems: these are mobile communication systems for the exchange of data, in which at least one hop of the communication path is accomplished by a wireless link. In this way, as we can observe, *system heterogeneity* becomes one significant characteristic of mobile P2P systems compared to conventional systems based on fixed networks.

4.3.1 Benefits for Mobile Users, Mobile Operators and Service Providers

With the increasing availability of data communications, including Internet access in mobile networks, P2P applications originating in wired network also become available to mobile users. They want to enjoy the same level of service while they are mobile as they have received from fixed line access.

This is just one reason why a new range of applications may appear in mobile networks. Another major reason is that P2P technology makes the development of certain applications easier and faster. To state this more precisely, once a P2P platform is provided (including routing, a data management layer, etc.) building new applications on top of it becomes easy; in many cases it only involves refining the provided data management functionality and selecting appropriate data abstractions to meet application and performance requirements. This is why mobile operators are interested in providing such P2P platforms as an incentive for third parties to develop new applications. These third-party service providers are currently not common in the mobile business and could provide a new value within it. Moreover, we believe that these third parties may be even the mobile users themselves. However, whether this is the case or not, a whole new market may open up, in which various services are offered by the users and traded among them. Services based on the local proximity of the users are an interesting example. Here, users who are close to one another could exchange different types of information, such as information about free carparks, good restaurants, places to visit within a certain area of a city, information about sales in local shops, etc.

Furthermore, the P2P paradigm allows operators to extend their service coverage to the ubiquitous environment (i.e. loosely coupled systems with different types of air interfaces) and systems that are not using centralized servers, such as today's cellular systems for service provisioning. A broad range of new applications using and processing data from sensor networks might emerge. Cellular phones can be used in such applications to provide a ubiquitous interface to users or as mobile sensors. In the next section we outline one scenario where this could be useful.

Last but not least, P2P systems do not rely on infrastructure but use users' equipment for service provisioning. Mobile communications based on available user equipment rather than on expensive infrastructure reduce the maintenance cost and allow much cheaper service offerings compared to conventional systems.

4.3.2 Mobile P2P Applications

In principle, many of the most conventional P2P applications can also be realized in mobile P2P systems to increase the number of users and the application portfolio in mobile networks.

Furthermore, the capabilities of mobile networks allow new kinds of P2P applications, such as location-based applications, formation of spontaneous communities and applications for ubiquitous environments, to name but a few.

In general, categories of P2P applications include sharing of digital information, personal communication, community and collaboration services, overlay routing indirection (for example multicast), network management, data streaming and grid-based services. In addition to well known or emerging P2P applications such as file sharing or voice over P2P, P2P mechanisms can also support general Internet services such as multicast, anycast and mobility through a P2P overlay network, introducing a new layer of indirection on top of the IP-layer.[5,35] Conventional network management can benefit from P2P techniques. P2P data streaming applications are different from the P2P data-sharing concept, where files are accessed after download. In P2P data streaming, media can already be played during download, creating a growing chain of streaming peers. Decentralized service provisioning is another application of P2P concepts, where, for example, web services can be leveraged from their client/server paradigm executed in a P2P grid. In the following, we will look in more detail at mobile P2P-related applications.

4.3.2.1 Information Sharing

File sharing is the most prominent example of P2P applications and enables sharing of digitized information among the users of a P2P system. However, it is just an example. This application type can be generalized to sharing any kind of user-authored information, including very popular applications such as web logging or blogging, where users publish their comments and information to the online world. We emphasize that, in a mobile environment, such P2P applications could be enhanced with location awareness and, as such, they could become even more useful and popular. Location awareness implies providing the most relevant information depending on the requester's location. Sensor networks in which information about the users and the environmental context are shared can also benefit from P2P query mechanisms.

4.3.2.2 Communities Formation and Group Communication

Based on location or simply on interest, P2P technology supports spontaneous formation of virtual communities without additional infrastructure (such as a presence server). Applications include simple information sharing, as described above, based on a certain topic (such as the world soccer championship), based on the location (a stadium or a concert) or based on some other shared characteristics (for example between cars in the same traffic jam). Services shared in spontaneous communities are not limited to information sharing, but include any kind of service, such as selling opera tickets before a performance or P2P gaming. Depending on the type of service, additional support might be needed concerning service negotiation and contract enforcement. The main building blocks for such applications are now available or are becoming so. For example, see Castro *et al.*[36] for a P2P group communication solution or Aekaterinidis and Triantafillou[37] for a content-based P2P publish–subscribe solution.

4.3.2.3 Personal Communication

It is not just since the recent success of Skype's voice over IP telephony (see http://www.skype.com) that P2P-based personal communication has gained interest (see http://www.p2psip.org). In addition to voice over IP, instant messaging applications use P2P systems to search for user addresses, replacing centralized registers. P2P technology saves cost since no infrastructure has to be maintained by the operators. Mobile applications again benefit, for example from location awareness and spontaneous, infrastructure-less community set up. Challenges include the provisioning of server-based cellular telephony features such as a voice box.

4.3.2.4 Ubiquitous Environments

As mentioned many times in the book so far, various devices have been enhanced through the addition of communication abilities. Examples are home networks, wireless sensor networks and telematics. In the near future, an environment will appear (a ubiquitous communication environment) in which many sensors, persons and different kinds of objects exist, move, and communicate with one another. We believe that P2P technology will play an important role in enabling useful applications in such an environment. To support this claim, imagine the following scenario. Customers of a shopping mall would like to have information about products such as their location, prices and availability, as well even as the opinions of other customers. Each shop could collect most of the necessary data via sensors installed in the shop. To make the information about their products globally (i.e. shopping-mall-wide) available, shops could run a small P2P network. The mobile users would be an integral part of the network. On entering the shopping mall they would join the network, use the service and finally leave the network. Technically, we would not expect the mobile users to participate actively in storing the available data or routing the requests because their possibly short online time might cause problems. Later in the chapter we outline a P2P architecture that meets such requirements.

4.3.3 Challenges of Mobile P2P

There are many challenges to solve in order to make mobile P2P a success. We give them below in the form of a list of performance requirements that we find particularly important (see Kellerer *et al.*[38] for further discussion on this). Note that the last requirement in the list, i.e. trust and incentive models, can be extremely important, not only for improving the performance of the selected P2P solution but also for the user acceptance of the application in question, because it is normally possible to use the same models for tuning user behaviour at the level of the application and at the level of P2P routing.

Requirements include:

- Reduce the P2P lookup traffic overhead as much as possible in order to overcome the low transmission data rates of mobile devices.
- Address high churn rates due to frequent joining and leaving of nodes.
- Consider the limited resources of mobile devices and the heterogeneity of nodes and their distinct device capabilities.

- Minimize the search traffic generated in the physical network by taking into account its properties.
- Provide trust and incentive models to support users' willingness to comply with the protocol rules.

The first part of this chapter has been devoted to discussing the first requirement in this list. The main message of that discussion is that DHTs offer considerably better performance than unstructured systems. That was the main reason for our favouring DHTs throughout the chapter. However, the final decision about what P2P architecture is most suitable for a given mobile setting cannot be made on the basis of the lookup costs alone. Instead, it should be made on the basis of careful evaluations of all of the other requirements in the above list. In the rest of the chapter we will discuss some existing solutions and ideas of how this can be done. However, before that, in the rest of this section we describe how strict these requirements are in various mobile networks, i.e. how the underlying mobile network characteristics influence them. As we will see, in some networks they are less severe while in others they are quite critical. We differentiate between cellular networks, hot spots, ad hoc networks and sensor networks, listed in this order according to increasing constraints. For technical details on the wireless transmission systems discussed in the following, please refer to Chapter 2.

Cellular networks are one-hop wireless networks connecting peers to the fixed Internet via a single wireless link. Mobility management makes mobility issues, such as handover, transparent to the application. In this way, data-rate restrictions, in terms of throughput and delay, and device resources are most significant. With, for example, 9.6 kbps for GSM and less than 50 kbps for GPRS, peers connected via second-generation systems have to be regarded as low-performance nodes. Third-generation systems such as UMTS in Europe provide up to 384 kbps on the downlink. However, the uplink capability is still restricted to 64 kbps in most systems deployed. Cellular handheld terminals are usually constrained in storage space, processing power and battery. Though they offer only limited display capabilities, most data formats such as audio and video can be processed. Furthermore, more powerful devices such as laptops or PDAs can be connected either using a cellular handheld as a relay station or using a cellular data card. Mobile phones are mostly switched on for reachability. However, churn rates may still be an issue because of the high data transmission cost in cellular networks, causing users to turn off their data applications. This holds particularly for 2G systems. Heterogeneity is also an issue because of the widespread usage of 2G and 3G systems.

Wireless data-network hot spots offer comparatively high data rates over a locally restricted area. Wireless LANs based on IEEE 802.11b with up to 10 Mbps are the most commonly deployed infrastructure today. Whereas terminal capabilities and data rates are sufficient for P2P systems, mobility is restricted to the area around a hot spot. In other words, mobility is an issue, i.e. user mobility results in a high churn rate.

Wireless LAN technology is also common in Mobile Ad Hoc Networks (MANETs). Here, the transport link between peers is accomplished by multi-hop wireless links using intermediate nodes as relays. Since nodes in MANETs are assumed to be mobile, connection breaks are common whenever adjacent nodes move out of their radio range, which leads to high churn rates. In addition, MANET routing is costly in terms of bandwidth. This problem can be made even worse by a P2P overlay network that does not take the physical network topology into account. Thus, proximity awareness is another extremely critical issue for P2P over mobile ad hoc networks. In various MANET settings, the capabilities of the nodes involved also vary, so that the limited resources may also become critical.

Ad hoc sensor networks can be regarded as systems with extremely limited device resources connected over MANETs, as described above. They are only suitable for special types of P2P application.

4.3.4 Mobile Peer-to-Peer Overlays for Heterogeneous Mobile Environments

Mobile environments are usually characterized by the heterogeneity of the devices involved. They range from mobile phones to PDAs or laptops and can be connected via diverse access technologies such as GPRS, Wireless LAN or UMTS. Existing DHT solutions mostly focus on so-called flat P2P designs, i.e. designs in which all participating nodes are supposed to share the workload evenly. But this workload can be simply too high for some low-perform-ance nodes, for example mobile phones. Then, either these nodes must be removed from the system or the overall system operation is in danger. This motivates the exploration of new DHT architectures in which he heterogeneity of the nodes is explicitly taken into account. There is another strong reason for investigating such DHT architectures. Normally, commu-nication costs are different for different nodes. For instance, communication is usually more costly for mobile phones than for PCs with a DSL connection. Intuitively, if nodes with higher communication costs perform less routing (i.e. incur lower costs) while deriving the same benefits from the participation in the system, then the operation of the system as a whole is more efficient.

As far as unstructured P2P solutions are concerned, heterogeneity is usually not a problem in unstructured P2P networks. Here, low-performing nodes may simply fail in routing search requests and highly mobile nodes can leave the system without impact on other nodes. However, we note that there are hierarchical unstructured solutions (Gnutella 0.6 for example[39]) dealing indirectly with heterogeneity. Their main focus is on reducing the search complexity through a hierarchy, but they can be easily tuned to address the node heterogeneity.

Let us now describe two examples of DHT architectures designed specifically to deal with the node heterogeneity. The first one is the Hybrid Chord Protocol,[40]; see Figure 4-3.

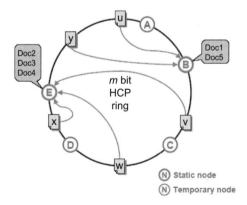

Figure 4-3.   Hybrid Chord Protocol

The Hybrid Chord Protocol distinguishes between two types of peers, static nodes and temporary nodes. Static nodes are highly available nodes with high-data-rate links, processing power and storage capacity. They are assumed to remain in the overlay network for a long time and thus store all the object references. Temporary nodes join the overlay network only for a short period of time. They do not store references to resources but they do perform routing. When a temporary node is responsible for a query, it forwards the query to the closest successor on the routing path (for example node D forwards a query for object W to node E). The Hybrid Chord Protocol uses a Chord ring as the basic query routing structure but can easily be adapted to any DHT.[41] Static nodes are chosen on the basis of uptime and performance. With this concept the background traffic in the DHT can be reduced significantly since high joining and leaving rates of temporary nodes do not cause shifting of resource keys. Furthermore, temporary nodes, (such as handhelds) are not loaded with storing and resolving keys.

However, this solution does not leave much flexibility for the nodes to optimize the network operation according to the cost model we hinted at above. Indeed, even performing routing may be too much for some nodes. In the above example this translates to moving the temporary nodes out of the Chord ring. Various hierarchical DHT designs can help here. Figure 4-4 shows an extremely simple DHT architecture in which extremely low-performance nodes are made free of DHT routing but still participate in the system. These nodes (call them leaves) are attached to their proxies (superpeers), which remain in the DHT. Zöls et al.[42] provide another interesting argument in favour of such a DHT architecture. The total operational cost of such a system is normally less than that of its flat equivalent, i.e. the system in which all nodes are moved into the DHT (provided the low-performance nodes can take the load in this case). The key to understand this is the observation that this hierarchical architecture actually presents a good trade-off between centralized systems, which minimize the operation costs but cannot be realized in practice because no single node can take the entire load generated in the system, and a totally decentralized system (a flat DHT), which incurs the highest operational costs and can be realized.

This simple hierarchical design is not the only one possible. Garcés-Erice et al.[21] mention a number of other possible hierarchical designs, for example one in which leaves operate within a new DHT instead of relying completely on their top-level DHT proxies. This dis-

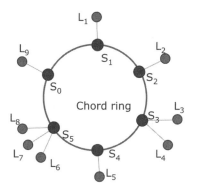

Figure 4-4. A hierarchical DHT

burdens the top-level nodes to a certain degree. In our opinion, it would be interesting to apply the cost model of Zöls *et al.*[42] to other hierarchical designs and compare their optimality across various settings.

4.3.5 P2P Overlay Networks and Awareness of the Physical Network Properties

It may seem from our discussions so far that the number of hops in the overlay network is the only metric used to measure the quality of the P2P object location algorithms. However, even though this metric is certainly very important, it is somewhat misleading. What matters in the end is the number of messages generated in the physical network, not he number in the overlay network. Thus, even though less traffic in the P2P overlay network normally leads to less traffic in the underlying physical network, it cannot be considered in isolation. Obviously, reducing the traffic in the physical network requires the overlay network to be aware of the physical network, its topology and other properties. Another downside of lack of awareness of the physical network is the search latency. Logically close overlay nodes should be as physically close to one another as possible. If this is not the case, the overall search latency may grow too large and seriously damage the system performance.

To illustrate how severe the problem can be, let us assume that we use an unstructured P2P overlay network on top of a MANET. To route a single overlay message, a node needs to discover the physical address of the destination node (depending on the MANET routing protocol, this can also require flooding the physical network). As a result, the total routing costs associated with the P2P overlay become extremely high, if not unacceptable. However, the problem can be eliminated if the overlay network is constructed based on the knowledge that the underlying network is a MANET. Long routes or zigzag routes in the P2P overlay network, which result from a lack of awareness of the underlying network topology, can be avoided if information from lower layers is taken into account., Gruber *et al.*[43,44] offer a simple and elegant solution applying cross-layer design principles. The proposed approach is called Mobile P2P (MPP) and is illustrated in Figure 4-5. The MPP uses the inter-layer communication specified in the MPCP protocol to insert queries into the network layer. In particular, the ad hoc network routing protocol, Dynamic Source Routing (DSR),[45] has been enhanced to allow the insertion of application layer P2P queries into Enhanced DSR (EDSR). As a result, queries are routed to the neighbouring node on the network layer instead of the application-layer neighbour, the proximity of which is unknown. At each node, queries are forwarded by the EDSR to the P2P application and checked against matching keywords. In the case of no success, the application layer inserts the query again for further forwarding. This leads to a more efficient search compared to a search on the overlay only.

The problem we have just seen is neither unique to unstructured P2P systems nor present only in MANETs. It is relevant in fixed networks and DHTs too, and the research community is aware of it. Currently there are two broad classes of solution aimed at reducing the physical network traffic and search latency, proximity neighbour selection and proximity route selection. The difference between them is rather subtle. In proximity neighbour selection routing tables are constructed based on proximity, while in proximity route selection the choice of the next hop depends on the proximity of the neighbours that are already in the routing table. Gummadi *et al.*[23] report that proximity neighbour selection outperforms proximity route selection. Thus, selecting peers carefully when constructing routing tables turns out to be

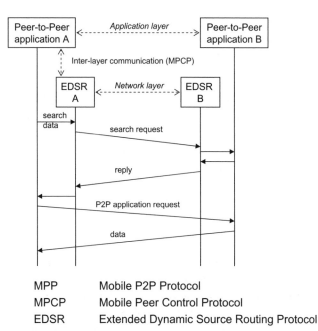

MPP   Mobile P2P Protocol
MPCP   Mobile Peer Control Protocol
EDSR   Extended Dynamic Source Routing Protocol

Figure 4-5.   Mobile peer-to-peer protocol

critical. Closely related to this is the flexibility with which the routing tables can be constructed – more flexibility certainly leads to a better choice of neighbours. Gummadi *et al.*[23]
further report that a satisfying level of flexibility is offered by ring- and tree-based routing
topologies for example (such as Chord and Pastry), while the hypercube topologies (such as
CAN) do not offer enough flexibility. It is interesting that the topologies offering constant
routing table sizes and logarithmic routing (such as Viceroy) do not offer any flexibility and,
as such, they do not admit proximity neighbour selection at all.

4.3.6 P2P Trust and Reputation Management

In any P2P application, peers provide services to one another. In order to adopt an application,
its users would normally want to make sure that the other peers will behave in a trustworthy
manner when provisioning the service. Reputation management has been recognized in the
research community as an appropriate way to boost trustworthy behaviour in P2P networks.
The key idea of P2P reputation systems is that collecting and disseminating information about
peers' past behaviour will influence their future business.[46] Here, a hidden assumption is that
the interactions among the peers are inherently repeated.

   P2P reputation systems have generated quite some interest, not only in the P2P research
community but also in the areas of artificial intelligence and economics. Most of the P2P
research on reputation systems (see, for example, Kamvar *et al.*[47] or Despotovic and Aberer[48])
assumes that the peers are characterized by static probability distributions that determine their
ability to perform in specific ways. The main focus of the proposed solutions is to discover
this distribution for any peer, based on the reports about its past performance, and make it

available to its interacting partner prior to their prospective interaction. Thus, the goal is just to signal what could go wrong in interactions and to what degree. It is unclear how precisely these signals can influence the future behaviour of the peers.

On the other hand, the concept of reputation has been a subject of study in game theory (see, for example, Kreps and Wilson[49]). The interaction between the involved parties is framed as a repeated game, implying that the parties (i.e. peers) behave as perfectly rational utility maximizers and are ultimately interested in maximizing their long-run payoffs (for the entire repeated game). The decisions about what action to take at any stage of the game are taken on the basis of the information about the outcomes of their or their partners' previous interactions. By selecting carefully in what form this information is made available to the interacting peers (feedback aggregation), it is possible to influence directly their future performances. Ideally, it is possible to select trustworthy behaviour (associated with specific actions in the game played) as their optimal strategy.

All this can be in principle be applied to P2P routing itself, not only to P2P applications that assume that a correctly operating P2P routing layer is providing whatever reputation data is required. The problem is, however, somewhat more complicated in the sense that there is now a mutual dependency between the reputation system and the core P2P routing layer: the former needs the latter for its proper operation and vice versa. We are not aware of any work evaluating such a system, i.e. answering how well the reputation system would work and what impacts on the routing it would have. See Blanc *et al.*[50] for more insight on how to frame a routing game and a discussion on the effects of reputation management.

Both probabilistic and rational models are somewhat extreme in our opinion. They are not fully applicable to settings in which humans participate because humans are not fully rational and their behaviour is not totally probabilistic. We believe that it is worth while investigating other forms of behaviour characterized by the ability to learn from the past but not fully optimize across the entire lifetime. Simple models involving various forms of reciprocation are a good example.

4.4 Summary and Conclusion

P2P systems are emerging as an important paradigm that can have a disruptive impact on the service platform. It can enable a whole new range of applications and bring the service platform to areas that it was hard to imagine just a few years ago. Most notably, ubiquitous environments merging sensor, cellular and fixed networks may benefit from the P2P paradigm.

In the first part of this chapter, we described the P2P concept and analysed the main classes of P2P systems, unstructured P2P networks and Distributed Hash Tables. We found DHTs more suitable for most of the settings, including those involving mobile participants. This is mainly due to their extreme scalability, i.e. the logarithmic complexity of the most important operations. However, we pointed out a number of difficulties related to the DHT concept and explained the overhead needed to overcome them.

In the second part, we discussed how to apply the P2P technology to mobile environments and listed the most important issues raised by this task. We found that most of the available DHT solutions, originally designed for fixed networks, cannot be transferred directly to mobile settings. Instead, new DHT architectures are needed that take into account the heterogeneity of the participating nodes, their limited resources and the constraints of the underlying physical network. Finally, we hinted at a number of proposed solutions that are taking steps in this direction. Even though these solutions demonstrate that the research community is making progress in providing mobile P2P platforms, we still believe that a lot is yet to be done in order to fully realize the potential of P2P technology in mobile settings.

5

Mobile Middleware

Chie Noda

Well-designed middleware provides valuable support to services and applications in a distributed environment. It conceals the heterogeneity of distributed systems, such as network protocols, operating systems and hardware. In particular, emerging ubiquitous networking and services in mobile communications raise the need for new middleware features running on heterogeneous devices. 'Mobile Middleware' is regarded as middleware specialized for a mobile network environment. It addresses the challenges of mobile, ubiquitous and resource-constrained devices.

Conventional cellular mobile communication (such as GSM, PDC and UMTS) is characterized by centralized service provisioning using gateway servers such as the i-mode service and the Multimedia Messaging Service (MMS). Emerging fixed wireless technologies for broadband (for example WLAN, WiFi and WiMAX) and Fixed Mobile Convergence (FMC) lead toward heterogeneity of multiple radio accesses provided by multiple operators and to roaming across multiple networks. Decentralization of access technologies may have the effect of creating an open and dynamically configured infrastructure that may not be addressed by centralized software systems. In such a dynamic, open and complex service environment, a user-centric paradigm driven by focusing on the demands and the behaviours of users is evolving as an important attribute of next-generation middleware.

With the recent development of mobile devices, we are experiencing increasingly powerful mobile and distributed computing environments. These phenomena foster the emergence of ubiquitous networking in mobile communication services. Mobile devices are expected to support short-range communication technologies (such as NFC and ZigBee) and act as mediators between conventional mobile networks and upcoming sensor networks. Integration of mobile devices and a wide range of networking technologies enables them to interact with the environment through ubiquitous devices augmenting everyday objects and surrounding users, to retrieve user environmental information (such as location and temperature) and user status (for example at work, on vacation, driving the car) and to further actuate assistance to users.[1]

Towards 4G Technologies. Edited by Hendrik Berndt.
© 2008 John Wiley & Sons, Ltd.

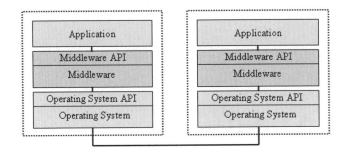

Figure 5-1. Middleware layer

Middleware is defined as a distributed platform providing interfaces and services to facilitate the development, deployment and management of distributed applications in heterogeneous environments. Middleware is a class of technologies to help manage complexity and heterogeneity inherent in distributed systems.[2] It lightens the burden of distributed system programmers by concealing heterogeneity, for example heterogeneous operating systems, hardware, network protocols and databases. It is designed as a middle layer of software between operating systems and distributed applications, as shown in Figure 5-1. Middleware connects parts of a distributed application with data pipes and exchanges data between them.

Mobile middleware is regarded as middleware specialized for a mobile network environment, to support mobility and, in particular, to run on resource-limited mobile devices. As also discussed by Gaddah and Kunz,[3] specific requirements for mobile middleware, which are usually not considered in conventional middleware platforms running on broadband wired networks and resource-rich devices, are as follows:

- *Light resource.* Mobile middleware is to be distributed to resource-limited devices with low performance, a limited amount of memory, narrow radio bandwidth and requirements of lower power consumption. Even if hardware evolution follows Moore's law, they will still have limitations on resources and capabilities compared with other nodes, such as network nodes and application servers. Thus, mobile middleware has to be designed to optimize light resources.
- *Asynchronous communication.* Mobile network connections are characterized by limited and varying data rates and frequent disconnections, for example caused by mobile devices moving out of network coverage or simply being shut off. Communication speed is expected to be much higher, and the downlink is assumed to have a much higher speed than the uplink. If these characteristics are taken into account, mobile middleware has to support asynchronous communication, with decoupling client and server, and provide reliability.
- *Context awareness and adaptation.* Quality of Service (QoS) will change dynamically as a result of the heterogeneity of radio access technologies. Not only QoS but also general application contexts, such as location, and the user context will be dynamic. Mobile middleware needs to support context awareness, i.e. be aware of context changes and adapt its behavior dynamically or forward this information to other parts of the system. For example, when the context information is provided to an upper layer to assist users (see details in Section 5.3) or a lower layer to optimize network parameters (see details in

Section 5.5), the system can support user-centric service provisioning and enhance the system performance.

Hereinafter we discuss attempts to build mobile middleware from four different aspects:

- mobile middleware technologies as an extension of conventional platforms to support mobility and resource-limitations
- functional components required in middleware architecture
- a pattern to make software components more interactive for service discovery and for negotiation toward service delivery
- evolution of mobile devices and examples of smart devices supporting mobile middleware

5.1 Mobile Middleware Technologies

Many research efforts have been focused on designing middleware platforms for mobile environments. We discuss four different types of middleware technologies (publish/subscribe, reflective, mobile agent and peer-to-peer middleware) and describe some of the developed solutions as examples. Each of them aims to support at least one of the above three requirements. We analyse their advantages and disadvantages for middleware adaptation to mobile environments, and extend this to resource-limited devices.

5.1.1 Publish/Subscribe Middleware

Publish/subscribe middleware, which is also known as event-based middleware, is a paradigm for building large-scale distributed systems. Senders and receivers are not required to have knowledge of each other in advance. Receivers subscribe to information that is of interest to them without specifying any specific sources, while senders simply publish information without addressing it to specific destinations. Publish/subscribe middleware interrupts subscriptions and delivers the relevant publications to subscribers.

There are standardized publish/subscribe middleware specifications, for example CORBA NS (Notification Service)[4] and JMS (Java Messaging Service).[5]

The publish/subscribe paradigm supports asynchronous communications and context awareness. Existing publish/subscribe middleware platforms have been extended to support the mobile context, for example a Java-based object-oriented middleware utilizing mobile agents.[6]

5.1.2 Reflective Middleware

The concept of reflection was introduced in the field of programming languages design as a principle that allows a program to access, reason about and alter its own interpretation.[7] It has been recently applied to distributed systems. Reflective middleware provides inspection and adaptation of systems' behaviours by utilizing meta-interfaces. It enables systems to become open, configurable and re-configurable. Reflective middleware is constructed as a group of collaborating components. It enables configuration of small middleware engines that are able to interoperate with other traditional middleware.

Dynamic TAO is a reflective component-based middleware platform that CORBA ORB developed as an extension of TAO.[8] The mobile reconfiguration agent is built on top of Dynamic TAO.[9] Another example is XMIDDLE[10] extended with mobile code techniques.[11]

5.1.3 Mobile Agent Middleware

Mobile code is executable program code that moves from a source machine to a target machine, where it is executed. It can improve speed, flexibility and ability to handle disconnections. It enables dynamic code installation in order to extend the functionalities of existing devices and systems, which may solve the problem of resource limitations.

A mobile agent moves into the execution environment together with the code and the data, such as the current state. It allows the upgrading of distributed objects without suspending services. In contrast to mobile code mobile agents navigate autonomously: they decide on their own whether and when they migrate. Mobile agents fit to mobile environments, characterized by limited available bandwidth and network disconnections, by supporting asynchronous communications. Network connectivity is required only during the period when the agent migrates from one place to another.

In addition, mobile agents can support dynamic adaptation and personalization according to the current context information. Mobile agents can support a certain level of intelligence, ranging from pre-defined policies to self-learning artificial intelligence mechanisms, and act autonomously. Agents can also coordinate with other agents to achieve a common goal and share the computational load. Mobile agents promise to cope more efficiently with a dynamic, heterogeneous and open environment.

The Java Virtual Machine and Java's class loading provide useful features for developing mobile agent middleware. The most important of these features are serialization, remote method invocation, multi-threading and reflection. Several mobile agent middleware platforms, implemented on Java and compliant with standards, either OMG's MASIF (Mobile Agent Systems Interoperability Facility) or FIPA (Foundation for Intelligent Physical Agents), are available on the market. On one hand, MASIF defines standard interfaces for the basic functions of agent management and transfer between heterogeneous mobile agent systems. On the other hand, FIPA standardizes the general architecture of agent platforms and focuses on interoperable agent communication languages. Mobility of agents is not mandatory in FIPA. As examples, IBM's Aglets[12] and SOMA[13] are MASIF compliant, whilst FIPA-OS,[14] JADE[15] and LEAP[16] are FIPA-based agent platforms.

5.1.4 Peer-to-Peer Middleware

There are basically two communication models for distributed applications: the client–server model and the peer-to-peer model. The former is a traditional request-and-response style, where applications components are divided into two types, those that make requests (clients) and those that fulfil requirements (servers). The latter is a conversational communication model between two applications that exchange information for service discovery. There is no longer a central point to publish services and information, and all participants can transparently share information in a global space. It enables context awareness in a peer-to-peer manner.

Table 5-1. Comparison of mobile middleware technologies

	Light resource	Asynchronous communications	Context awareness
Publish/subscribe middleware		X	X
Reflective middleware	X		X
Mobile agent middleware	X	X	
Peer-to-peer middleware	X		X

Characterized by the ability to create, join and interact with peer groups, and to post advertisements, peer-to-peer middleware can dynamically find needs, adapt to changes and collaborate among peers to achieve a common goal of a community in an ad-hoc manner. It also enables distribution of the computational load across multiple peers and sharing resources.

Dynamic service discovery is a key technology for peer-to-peer middleware, such as JXTA,[17] SDO (Super Distributed Objects)[18] at OMG (Object Management Group) and Apple Computer's Bonjour[19] as an implementation of the IETF zero-configuration protocol.[20]

5.1.5 Challenges for Mobile Middleware Platforms

The approaches of publish/subscribe and reflective middleware are to rebuild a traditional middleware to fulfil the requirements of mobile middleware. The paradigm of mobile agents can be applied to it as an additional service and peer-to-peer middleware as an underlying communication mechanism. Table 5-1 shows the advantages of each mobile middleware approach.

The configurability of reflective middleware, the code mobility and installation of mobile agent middleware and the distributed computational load of peer-to-peer middleware are all designed to support resource-limited devices. Note that some of the above-mentioned middleware platforms require more resources than those that current mobile devices support. Enhancement of device capabilities may solve this problem.

However, none of these approaches fulfils all the requirements for mobile middleware. Some research has integrated one platform with another technology that complements it, for example publish/subscribe middleware with P2P communication[21] or reflective middleware with mobile agents.[11,22] We anticipate that the integration of these technologies to fulfil the requirements for mobile systems is of major importance to achieving the goal.

5.2 Architecture Components

In this section, we discuss the functional components required in middleware platforms to support new capabilities for next-generation networks and the service architectures described in Chapters 2 and 3. Figure 5-2 shows a logical view of middleware architecture.

The network support layer provides functionalities for network communications control in heterogeneous networks, as described in Section 2.2. The user support layer supports the autonomous and proactive agents that traditional service middleware lacks. It enables provi-

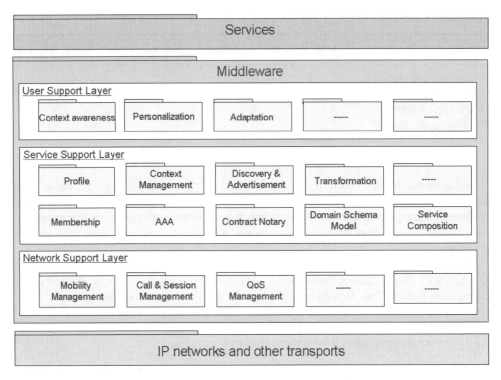

Figure 5-2. Middleware logical architecture

sion of user-centric services, based on context awareness, personalization and adaptation, as described in Section 3.2. Hereinafter we describe each functional component in the service support layer.

5.2.1 Service Support Layer

5.2.1.1 Profile

Profiles are collections of data that may be used to adapt services to a user's specific environment and preferences, i.e. personalization. Profiles can be classified in the following categories: *user profile*, *service profile*, *terminal profile* and *network profile*. This middleware functional component supports management of profiles, by offering a means to select, retrieve, store and update profiles to users. A user may have more than one profile and select the most appropriate one, or it may be chosen automatically on the basis of other criteria. (For more details see Chapter 7.)

5.2.1.2 Context Management

Context is a set of environmental states which either determines an application's behaviour or in which an application event occurs. Contexts can be classified in the following four categories: *user context* (for example the user's location and the current social situation); *time context*; *physical context* (for example lighting, traffic condition, temperature); and *computing context* (for example communication data rate, available memory and processing power, battery status). This component offers a means to provide context information to services. It has a mechanism to sense and retrieve the current context, for example via sensors, and deliver it to other services. It also records a context history across a time span. Moreover, it monitors context changes and sends these to others using the subscribe/publish interface. (For more details see Chapter 9.)

5.2.1.3 Discovery and Advertisement

This is used to dynamically publish and search for information describing services or resources. It also provides a means to register services and dynamically required resources in the registry, update them and delete them from the registry.

5.2.1.4 Transformation

This supports adaptation of content that is processed and produced by services. There are several aspects that can be transformed: *format and encoding* (for example GIF and JPEG files); *type* (such as video, audio and text); and *structure* (for example XML and HTML).

5.2.1.5 Membership

This provides a way of managing groups and their members. Groups can be used to easily specify and address groups of entities that share common interests and goals. It also helps to manage the dynamic changes involved in a group's formation, such as adding a new member, removing a member and managing the life cycle of membership.

5.2.1.6 Authorization, Authentication and Accounting (AAA)

This is used to identify and authenticate principals, to verify and to grant authority by enforcing access control based on a set of authorization policy rules. Accounting is aimed at processing and managing accounting-related information, by calculating usage costs in order to charge users for making use of the services and resources provided.

5.2.1.7 Contract Notary

This is used to manage and validate contracts. Contracts are stored persistently in a repository. Contract notary also supports retrieval, registry, deletion, updating and validation of contracts in the repository.

5.2.1.8 Domain Schema Model

The domain schema model accepts a given set of structured data and returns artefacts of the model in the format of a schema, or ontology, which has been produced by a domain expert. A schema or ontology can range from a simple taxonomy with semantically weak relationships to a conceptual model with more complex link semantics and specification. Domain experts, using a defined specification language suitable for modelling links and associations, deposit domain model artefacts into the component. (For more detail see Chapter 7.)

5.2.1.9 Service Composition

The purpose of service composition is to create and dynamically provide at run time new services from a set of service components. It enables the integration of existing services to satisfy complex requests as a result of cooperation between multiple services.

5.3 Dynamic Service Delivery Pattern

The dynamic service delivery pattern is an example of configuration being used for the inter-working of the middleware functional components discussed in Section 5.2. It supports service discovery and autonomous negotiation leading to service agreement and delivery in heterogeneous environments without a central point, where services will be registered or can be discovered. Entities can be any piece of software or any software component. They play one or more roles, such as *Participant, Coordinator* (in the negotiation phase), *User* and *Service Provider* (in the service delivery phase).

The dynamic service delivery pattern consists of three phases:

1 *Introduction phase.* the entities interact with each other by using the publish/subscribe interface for service discovery and advertisement.
2 *Negotiation phase.* The same entities act as *Participants* and a trusted entity is also involved, playing the role of *Coordinator*. The participants exchange negotiation schemes, credentials and proposals in order to reach a service agreement according to the scheme agreed by the coordinator.
3 *Service delivery phase.* The entities acting as *User* and *Service Provider* interact to fulfil the terms of the service agreement. A *Service Session Manager* manages the service delivery and terminates the service session upon completion of the service delivery.

Figure 5-3 depicts an example of entities and their roles. The entities A and B, which share a common interest for service offering, take the role of user and service provider respectively. Both entities take roles of participants during the negotiation phase. The entity B further acts as a coordinator, who processes negotiation interactions between the participants according to the negotiation schema. Eventually the entity B delivers the service to A during the service delivery phase.

Figure 5-4 shows the sequence of the dynamic service delivery pattern according to the role assignment in Figure 5-3. The negotiation starts by creating a coordinator. Agreeing a *negotiation schema*, which defines rules and protocols on how the *coordinator* should carry out negotiation processes, is optional. The coordinator can pre-define the *negotiation schema*. The *coordinator* asks the *participants* to submit initial proposals, such as an offer and a demand, based on a participant's *negotiation strategy*. It checks if the two demands match; if not, it creates modified proposals according to the *negotiation schema* and sends them back

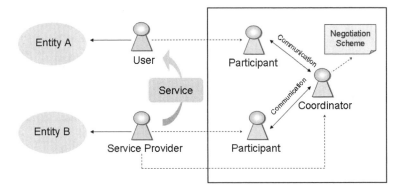

Figure 5-3. Example of entities' role for the dynamic service delivery pattern

to the *participants*. The *participants* check if they are acceptable according to their *negotiation strategies*. Once a *service agreement* is achieved, the coordinator files it as a contract and creates a *service session manager* to control service sessions. The *service session manager* monitors transactions between participants' entities to enforce service delivery.

The pattern is used whenever an entity needs to interact with another. It allows the system to support re-configurability and adaptability at both middleware and application/service levels, by supporting flexible internal interactions within the system, as well as external interactions.

5.4 Smart Devices Supporting Mobile Middleware

For the last half decade, we have seen two major streams of mobile-device evolution. One is in the direction of highly capable mobile devices, which are used mainly for human-to-human or human-to-machine communications. Most of them support Internet browsers, cameras and audio/video interfaces. Some also support a credit-card function by integrating smart-card technologies. We call them 'All-in-One' devices, which bring the environment users need into a single mobile device. This is on the same track as current high-end mobile-device evolution.

Another direction, the 'Device-in-All', is for machine-centric communications. Data-specific mobile terminals supporting additional radio interfaces, such as GPRS, WLAN, WiFi and WiMAX, which can be plugged into PCs, are examples in the current market. In the future, the evolution of device hardware will enable miniaturization of a the 'Device-in-All' into a tiny single chip, the so-called 'Chip-in-All'. They will be embedded into or attached to many kinds of device and everyday objects allowing active interactions with the user's environment. We foresee that 'Chip-in-All' will become the main stream towards ubiquity, while 'All-in-One' devices will coexist with them as the principal personal mobile devices (Figure 5-5).

In addition to mobile devices, there is another type of device in users' hands. Smart cards, the so called 'SIM's, are used as a personal and secure device in the widely deployed GSM and the recently introduced UMTS mobile systems. They provide a direct linkage between operators and users by storing subscriber identities and supporting network authentication.

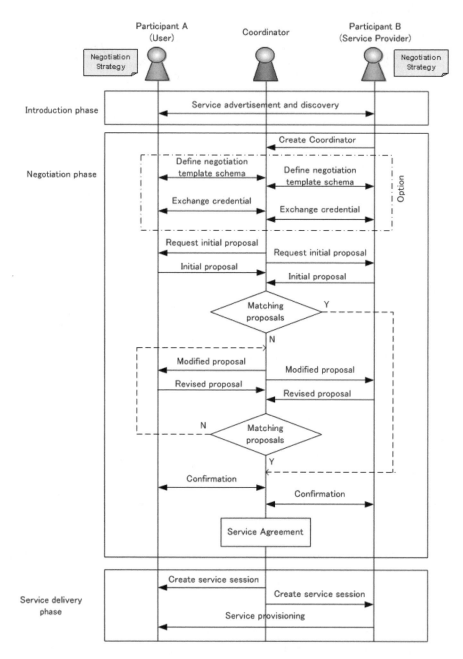

Figure 5-4.   Sequence of the dynamic service delivery pattern

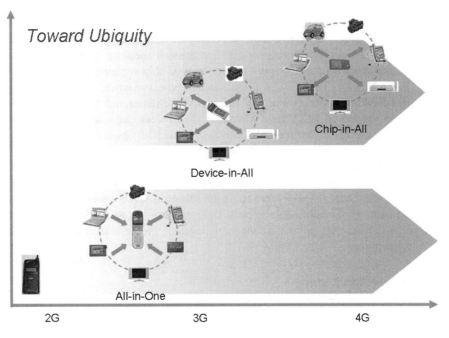

Figure 5-5.   Evolution of mobile devices

Keeping this evolution scenario of mobile devices in mind, we envision that the next generation of smart cards will play a key role as a meeting point with users and service providers.
In contrast to most of the ubiquitous devices that may be invisible to users and embedded or
integrated to everyday objects, smart devices are regarded as personal and visible devices,
providing secure execution environments for application tasks and storage spaces for privacy
concerning sensitive information.

Hereinafter, we look at the state-of-the-art of smart-card related technologies and discuss
the following future trends:

- smart cards for mobile telecommunication, such as SIM/UICC, to store subscriber information and further support an open execution environment
- contactless smart cards embedded into mobile phones, as a step toward ubiquity
- RFID, for which mobile devices support a reader function
- the evolution of smart devices, playing a role similar to smart cards in a ubiquitous
  environment.

5.4.1 Smart Card Technologies

Smart cards, the so-called SIM (Subscriber Identity Module) and 3G UICC (Universal
Integrated Circuit Card) cards, are globally used today in both GSM and UMTS networks.

5.4.1.1 Physical Architecture

Smart cards, often called integrated circuit (IC) cards, are defined as an IC chip embedded in a plastic card. They contain components for data transmission, memory storage and processing.

The key component of smart cards is a tamperproof IC chip containing the communication interface. In the 1990s an 8-bit CPU was the standard type, but nowadays a 32-bit CPU with a 30–50 MHz clock rate is becoming popular to meet the requirement of high performance, for example to support a Java Virtual Machine. Most recent chips are also equipped with dedicated co-processors for cryptographic functions, which results in much higher computation performance than with software coding.

Smart cards contain both volatile memory and non-volatile memory. For fast read and write access, new memory, such as FeRAM (Ferroelectrics Random Access Memory) and MRAM (Magnetic Random Access Memory) are being introduced as less expensive technologies than today's flash memory. The chips of the early 1990s offered less than 10 Kbytes of memory, today chips with memories up to 1 Mbyte have appeared on the market and these are used for multi-applications.

5.4.1.2 Software Architecture

The software architecture of smart cards is based on a file system and associated access commands specified by ISO7816-4. Security commands such as authentication or pin verification are usually customized and expanded in industrial standards such as ETSI SCP or 3GPP for telecommunication SIM/USIM, and EMV[25] for payment systems.

The OS of the smart card provides basic functions, such as communication, memory control and cryptography. One of the main roles for the card OS is to control the electrical and physical interfaces of the smart-card hardware. A Virtual Machine (VM) provides the environment that enables multi-application download and execution. There are industrial standards for OSs and VMs, such as Java Card[26] and MULTOS.[27] MULTOS covers not only VM but also application certification procedures and other operational aspects, because it was originally specified for loading credit applications onto smart cards (for MasterCard). Java Card is the most popular application-independent platform. Sun Microsystems defines the minimum subset of the Java specification as Java Card. Java Card specifies the tiny, but Java-compatible, virtual machine and runtime environment including smart-card-dedicated APIs. Application developers develop the Java Applet, which includes the file structure and commands. Figure 5-6 depicts the architecture of Java Card.

3GPP and ETSI SCP specify the SIM/USIM Application Toolkit (SAT/USAT) for SIM/ UICC to make them act proactively, rather than as a slave of the external reader. SAT/USAT APIs are designed to facilitate the development and downloading of SAT/USAT applications over the air or through Bluetooth. SAT/USAT provides a mechanism that allows applications stored on SIM/UICC to interact and operate with mobile terminals, providing operations such as displaying text, sending SMS messages and launching the browser corresponding to a URL.

5.4.2 Contactless Smart-Card Technologies

Another type of smart card in terms of communication mechanisms is contactless, as it does not require metal interfaces. These are popular in the public transportation domain, for

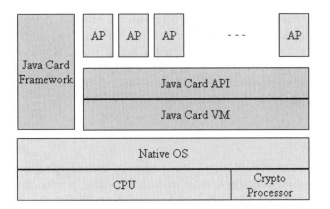

Figure 5-6. Java card architecture

example for charging transportation expenses and storing commuter tickets; (the Octopus Card in Hong Kong,[28] the Oyster Card in London[29] and the Navigo pass in Paris[30]). Contactless smart cards get their power supply from external readers through wireless antennas, which generate electromagnetic fields. ISO/IEC 10536, ISO/IEC 14443, and ISO/IEC 15693 specify different types of communication protocols depending on the distance. ISO/IEC 10536 is called closed coupling, which means that the communication distance allowed is very short such that the card should be used attached to the terminal, with a 4.91 MHz frequency. Cards conforming to ISO/IEC 14443 are called proximity cards and allow a communication distance of 0.2 mm to 10 cm, with a 13.56 MHz frequency. ISO/IEC 15693 cards are called vicinity cards and are enabled for a communication distance in the range of 70 cm with a 13.56 MHz frequency.

One example of widely deployed contactless smart cards in mobile communication systems is that NTT DoCoMo's i-mode FeliCa 'Mobile Wallet' service launched in July 2004.[31] In 2007, more than twenty million mobile terminals equipped with FeliCa were available on the market, and more than 22 thousand shops provided services for them, such as e-money and e-ticketing. FeliCa is a contactless smart-card technology developed by Sony.[32] The contactless interface communications are compatible with the ISO/IEC 18092 standard, proposed by Sony and Philips as Near Field Communication (NFC). They support a communication speed of 212 Kbps with clock rate of 13.56 MHz. All processing, from card detection, through mutual authentication, to data read/write, is completed in 0.1 second. FeliCa supports READ/WRITE commands as well as specific commands for e-commerce applications. The currently available FeliCa cards support 8-bit RISC CPU and 2.0 to 4.0 Kbytes for user/application memory.

In i-mode FeliCa, the FeliCa chip and a loop antenna for external contactless communications are embedded into the mobile devices. FeliCa chips can operate by detecting weak electronic signals emitted by a reader/writer (an external device). FeliCa applications are further integrated with i-appli on mobile devices, which are based on Java applications for J2ME CLDC (Connected Limited Device Configuration) and KVM (K Virtual Machine).[33] This allows on-line FeliCa application download and update (such as value charge), as well as local access to data stored on FeliCa (for example the mobile phone as a viewer of the value).

i-mode FeliCa is used for e-commerce, e-money, e-tickets and storage of identities such as membership or employee status. In 2006 East Japan Railway Company (JR East) announced the launch of a service based on i-mode FeliCa, which would allow users to use mobile phones as commuter passes and to recharge the stored fares. This phenomenon is likely to lead to an increase of machine-centric communications and interactions with the environment as a step toward ubiquity, in the form of 'All-in-One' devices.

5.4.3 RFID Technologies

RFID (Radio Frequency Identification) is a technology that incorporates the use of electromagnetic or electrostatic coupling at radio frequency in order to uniquely identify an object, animal or person by storing data. RFIDs, which are also called RF tags or smart labels, are used, for example, for logistics, distribution, supply chain applications.[34] Unlike barcodes, RFIDs are resistant to moisture or heat during the manufacturing or distribution process, by enabling automatic identification without physical contact. The read range can be extended depending on the type of RFID used, from a few inches to hundreds of feet by applying different frequencies, as specified in ISO/IEC 18000. This standard specifies radio interfaces and protocols, but not a physical format for RFIDs. Thus, RFIDs come in a variety of different shapes and sizes such as plastic cards, stickers, wristbands, coins, labels, etc. RFIDs can be powered internally, offering further flexibility, which also differs from smart cards.

RFIDs only communicate when they are in range of external readers. Their durability is obvious through their long life span. RF tags have two types of proactiveness:

- Active RF tags are powered by an internal battery as well as from external reader devices. Active RF tags are more expensive and larger than passive RF tags. However, they are also more powerful and have a larger reading range.
- Passive RF tags are powered by the field generated by external readers. Passive tags are typically much lighter than active tags, they are less expensive and they offer a virtually unlimited operational lifetime. However they have shorter reading ranges and require a higher-powered reader than active tags.

Most of the current RFIDs are implemented as memory cards at a lower price, without the capability of carrying out computation. The information stored on RFIDs is simply read by external readers. One can assume that RFID can be enhanced by supporting additional capabilities such as computation and tamper resistance, like smart cards. The boundary between contactless smart cards and RFIDs is coming closer and closer. Mobile terminals that act as RFID readers or support NFC interfaces are already on the market.[35]

5.4.4 Examples of Smart Devices

New application scenarios for nomadic users equipped with smart devices are emerging; these, on one hand, allow constant interaction with the users' environment through ubiquitous devices, while, on the other, they require a high level of context awareness and software adaptability to any changes. In the future smart devices will play the same role that smart cards do at present for mobile systems, providing identities for network security and a meeting point between operators and users. It is anticipated that they will further integrate short-range local communications, such as NFC, RFID or ZigBee, so as to be able to act as a mediator between conventional mobile networks and sensor networks. This will allow smart devices to be integrated into ubiquitous communications. Smart devices can interact with ubiquitous devices attached to everyday objects and could also make up sensor networks, supporting a much wider range of applications and services. They may also act as a gateway in local ad-hoc networks, providing value-added services in ubiquitous environments.

To support context awareness and software adaptability, smart devices will be incorporated into middleware platforms. With the expected evolution of device hardware and the development of open software environments, they are likely be a target for distribution of software. However, in comparison with other nodes, such as 'All-in-One' types of mobile devices, peripheral devices (such as home appliances), application servers and gateways, they still have limited resources and capabilities. Only fundamental software functionalities will be distributed to smart devices. Adaptation of middleware to smart devices will enable fast deployment of services without consideration of the hardware or the communication layers.

For example, smart devices can be one of the most appropriate places for users' personal software agents to reside. Users will always be able to carry their own personal agents on their smart devices and then let them migrate to other places and execute tasks autonomously but securely, such as negotiation with other agents, in both global and local networks.

Such an approach can be implemented on mobile agent middleware with an extension to support JXTA.[36] One of the advantages of hosting an agent software component on a smart device is its availability for local and ad-hoc networks, where a node is not always accessible to or reachable by smart devices. Another advantage is security and privacy, for example keeping personal sensitive information and credentials locally and establishing a federation of agents by utilizing cryptographic functionality.[37]

Another possible approach is given by Blefari-Melazzi *et al.*[38] A smart device has been introduced as a 'Simplicity Device', supporting users' network access and service usage for heterogeneous mobile and fixed systems. The Simplicity system consists of three main components, a smart device, a terminal broker and a network broker. The smart device is plugged into different terminals. It stores user's profiles, preferences and policies. The terminal brokers and the network brokers are based on a distributed broker architecture, which supports discovery, advertisement, adaptation and orchestration. The smart device allows the enforcement of personalized mechanisms to exploit service fruition, to drive automatic adaptation to terminal capabilities and to facilitate service adaptation to various network technologies and capabilities.

5.5 Summary

Emerging ubiquitous networking and services in mobile communications bring the need for new middleware solutions that can run on heterogeneous devices. This need has been recognized and mobile middleware is emerging as specialized middleware to address the challenges of mobile, ubiquitous and resource-constrained devices. In this chapter we discussed the current efforts to build mobile middleware. These include extending conventional platforms to support mobility and resource limitations, building new functional components, efforts to make software components more appropriate for service discovery and service delivery negotiation. We also discussed the present evolution of mobile devices, including smart cards, and gave examples of smart devices supporting mobile middleware.

6

Cross-Layer Design – a New Paradigm for Optimization of Mobile Communication Systems

Marco Sgroi and Wolfgang Kellerer

Cross-Layer Design (CLD) is a new paradigm that allows the optimization of communication network architectures across the traditional layer boundaries, but does not imply a change of the present layered architecture. In this chapter we describe the principles of CLD and show their application to the design and optimization of a wireless video streaming system.

6.1 Introduction

Next-generation wireless networks will have to support complex and resource-demanding applications, such as videoconferencing, 3D navigation and interactive gaming. Network operators will face the challenge of allocating the wireless medium efficiently to increase network capacity and provide services at the highest quality level to the largest possible number of users. The resource allocation problem is difficult to solve, especially because of the time-varying transmission characteristics of the wireless channel and the dynamically changing QoS requirements of most applications. Configuring a network statically and for the worst-case scenarios would lead to poor performance and inefficient utilization of resources. Instead, networks should dynamically adapt their configuration to the behaviour of the environment. Within the environment, both the application context and the wireless channel conditions are addressed. Dynamic adaptation requires a timely exchange of information across layers and a periodic reconfiguration of protocol layer parameters during network operation.

Network architectures have been traditionally designed following the layering paradigm. A layer is a group of communication functions that operate at the same level of abstraction,

Towards 4G Technologies. Edited by Hendrik Berndt.
© 2008 John Wiley & Sons, Ltd.

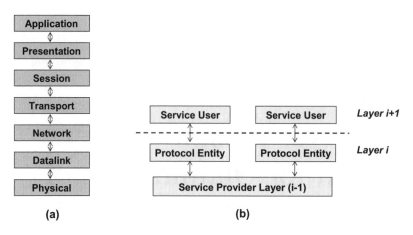

Figure 6-1. (a) OSI reference model; (b) Layered architecture

i.e. have similar rate of operation or handle packets of the same size. Hence, the communication functions of a network can be organized into a stack of layers, where each layer provides services to the upper layers and uses services offered by the lower layers. A common instance of the layering paradigm is the OSI Reference Model (Figure 6-1).[1]

The layering paradigm has been commonly used to design network architectures because:

- It applies the principle of separation of concerns and simplifies the design task.
- It favours modularity and allows replacing individual layers without changing the whole stack.

However, pure layering is no longer adequate to meet the challenge of designing next-generation mobile systems. The research community has shown several cases where architectures that are strictly layered have worse performance than architectures where multiple layers are jointly optimized. These cases mainly fall in two categories:

1 Layers are optimized locally and with respect to different metrics. Optimizing protocol layers locally and independently for different metrics may result in unintended behaviour of the network as a whole. One case of adverse interaction between layers is described by Kawadia and Kumar:[2] the network layer minimizes the number of hops to optimize delay, while the MAC layer adapts the data transmission rate to the received signal strength. Choosing routes with few long hops at the network layer results in low-data-rate links and overall in low throughput. The reason is that network and MAC layers are optimized independently with respect to different metrics (delay in the network layer, throughput in the MAC layer). The network layer ignores the throughput requirements of the application and selects a minimum-delay route without considering the throughput capabilities of the links.

2 Layers ignore the semantics of the messages originated in other layers. The provisioning of high QoS requires that layers are aware of the semantics of the packets they receive from other layers. For example, a layer should be able to identify the priority of the packets

coming from the upper layers and discard the least important ones in case the network is congested. Understanding the semantics of the messages originated in the lower layers allows a layer to better adapt to the dynamic variations of the physical medium. For example, the TCP protocol is based on the assumption that the underlying medium is reliable and that losses due to noise are negligible compared to those due to buffer congestion. This assumption is not always valid for a wireless medium. Therefore, when TCP runs over a wireless channel, the packet losses due to the high noise and interference are misinterpreted by the protocol, which reduces the transmission rate to avoid network congestion. As a result, the throughput of TCP over a wireless link is rather low. In this case the problem is not the layering approach itself, but that TCP is not able to distinguish between the different types of losses. Mechanisms of explicit congestion notification have been proposed to solve this problem.[3]

Video streaming offers another example of layers being not aware of message semantics. Usually a video stream consists of a sequence of video frames, which are partly dependent on each other for decoding and thus are of different importance for error-free display. Currently, the resource scheduling of a wireless transmission system does not take into account application parameters such as the different importance of independent and dependent frames and the resulting distortion when a frame is lost. We use this example to illustrate the principle and potential of cross-layer design in wireless networks in Sections 6.4 and 6.5.

To overcome the drawbacks of purely layered architectures, a new paradigm called Cross-Layer Design (CLD) has been recently proposed. CLD takes into account the dependencies and the interactions among layers and allows optimization across their boundaries. A common misconception about CLD is that it consists of designing networks without layers. Layering is just an *artefact* that allows simplifying the network design and management tasks. Within an architecture with identifiable layers of abstractions, CLD allows the joint optimization of the parameters of multiple layers. Therefore, CLD should not be viewed as an alternative to the layering paradigm, but rather as a *complement*. Layering and cross-layer optimization are tools that can be used together to design highly adaptive wireless networks.

Clark and Tennenhouse were among the first to warn against the loss of efficiency due to a too strict application of the layering principle.[4] They proposed Application Level Framing and Integrated Layer Processing (ILP) as a technique for optimizing protocol stack implementations across layers.

Recently, CLD has been applied mainly at the functional level to jointly optimize parameters of multiple layers. Haas argues that one of the major shortcomings of the layered architecture models is the lack of information sharing across layers, which prevents the network from quickly adapting to the changes of the environment.[5] Shakkottai *et al.* stress the importance of taking into account, up to the highest layers of the stack, the characteristics of the wireless medium that vary over time due to user mobility.[3] If multiple users share the physical medium, multi-user diversity gain can be achieved by allocating the channel resources to the users having a higher probability of successful transmission. Consider, for example, a system set-up including several mobile users and a base station as illustrated in Figure 6-2. The base-station scheduler dynamically allocates time slots to the users, taking into account the state of the channel for each user (for example expressed in terms of the transmission and error rates).

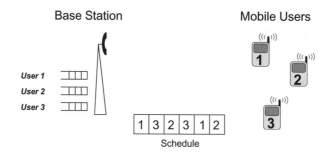

Figure 6-2.   Multi-user channel scheduling

Kawadia and Kumar take a rather different perspective on CLD.[2] First, they stress the importance of modularity and reconfigurability, arguing that they are better supported by layered architectures. Then, they point out that most papers on CLD do not consider the global effect of cross-layer optimization when multiple cross-layer interactions take place and network parameters are affected by conflicting adaptation loops. Rappaport *et al.* distinguish between cross-layer design and adaptivity.[6] The former is defined as a static optimization across the boundaries of the traditional layers and the latter as an optimization that takes place dynamically upon changes of the wireless medium or of the application requirements. Adve *et al.* present a framework in which layers dynamically adapt and cooperate towards a globally optimal configuration.[7] The assignment of resources is controlled by a central resource manager, which has access to global information. The resource manager examines the requirements of the application in the context of the state of the entire system and selects the configuration that maximizes the overall system utility and satisfies the resource constraints. Local adaptations are possible within specific layers as long as the allocated resources are not exceeded. For this chapter, we refer to CLD as the dynamic adaptation of several different layers of a protocol stack to optimize a certain global goal, such as the user-perceived Quality of Service. This optimization is based on the Cross-Layer Functional Architecture described in Section 6.2.

Previous work has applied CLD techniques to optimization of TCP over wireless,[8] cellular network channel scheduling,[9] ad hoc and sensor network protocol design[10,11] and wireless video streaming.[12–15]

Ludwig proposes a top-down approach to improve TCP performance over wireless,[8] where QoS requirements are passed from the transport to the datalink layer. Sternad used channel predictions to schedule wireless medium resources in time (time slots) and frequency (OFDM carriers).[9] Xylomenos and Polyzos defined a link layer that offers differentiated services and therefore can satisfy the requirements of multiple and diverse applications.[10]

The potential of applying CLD techniques to sensor networks is especially high.[17] Sensors are deeply embedded in the environment; therefore, networks are greatly affected by external changes, such as energy depletion, node mobility and variable interference, and must rapidly adapt to them. In the sensor networks domain CLD has been used mainly to jointly optimize:

• the network and MAC layers, to combine route selection with the node scheduling policy

- the middleware and network layers, to adapt the middleware to the variations of node connectivity[10]

The application of CLD to video streaming is discussed in detail in Section 4.

Although CLD techniques have already been applied successfully to the design of protocol stacks in special cases, further research is needed to answer more fundamental questions. In the following sections we lay out the basic principles for CLD together with the following fundamental questions, which are illustrated in an example in Section 6.5:

- When should cross-layer optimization be applied, taking into account performance gain and implementation cost? (See Section 6.3.)
- How should cross-layer optimization be implemented? As a central unit or as a set of components distributed across multiple network nodes and layers? (See Section 6.3.)
- Which parameter abstractions give the optimizer an accurate yet manageable description of the state of the layers? In order to be able to compute optimal settings of the system layers in real time, only a fraction of all available parameters characterizing system layers can be taken into account (see Section 6.2). Moreover, parameters have to be pre-computed (abstracted) so that they can be fed more easily into an optimizer.
- What metric should be used to optimize a, especially when it supports multiple concurrent applications of different types? Optimal user perceived quality of service for the worst-performing user might be one metric to optimize the system, while a maximum number of users, given a certain quality threshold, might be another metric.

6.2 Cross-Layer Functional Architecture

A Cross-Layer Architecture (CLA) includes multiple protocol layers and a Cross-Layer Optimizer (CLO). The CLO jointly optimizes multiple protocol layers of a network, taking abstractions of their predicted state and finding optimal values of their parameters. Figure 6-3 visualizes the components of a CLA.

Cross-layer optimization mainly consists of three steps:

1 *Layer Abstraction*, in which an abstraction of the predicted state of the layers is computed.

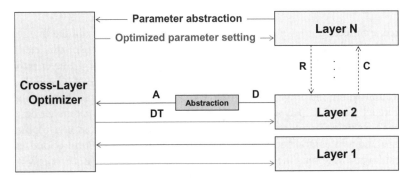

Figure 6-3.   Cross-layer architecture

2 *Optimization*, in which the values of the layer parameters that optimize a specific objective function are found.

3 *Layer Reconfiguration*, in which the optimal values of the parameters are distributed to the corresponding layers.

These steps are periodically repeated at a rate that should be set based on how fast the variations of the application requirements are and on what the transmission capabilities of the physical medium are.

Identifying a proper set of parameters that describe the state of the protocol layers is important. A description consisting of many parameters can be accurate but usually results in high communication overhead (to gather all the parameters from the network before optimization) and computation overhead (to explore the parameter space during optimization). Therefore, abstractions are used to reduce the number of parameters that are given to the CLO. Selecting the layer parameters to be jointly optimized is also an important step. While some layer parameters can be set directly by the CLO, others cannot be set directly, but change after other parameters have been set. Protocol layer parameters can be classified as follows:

- *Directly Tuneable (DT) parameters*. These can be set directly by the CLO. Example: time-slot assignment in a TDMA system or carrier assignment in an OFDM system.
- *Indirectly Tuneable (IT) parameters*. These cannot be set directly by the CLO, but may change after the DT parameters have been set. Example: bit-error rate that depends on the type of coding and the modulation scheme adopted.
- *Descriptive (D) parameters*. These can be read by the CLO, but not tuned. Example: frame rate or picture size in streaming video applications, which are set at encoding time.
- *Abstracted (A) parameters*. These are abstractions of descriptive parameters. Example: frame-loss probabilities that are derived from the channel state transition probabilities of the Gilbert–Elliott model, as shown by Peng et al.[12]

Layer parameters can also be classified along another dimension, with respect to the interaction among the network layers.

- *Capability (C)* parameters define the properties of a network layer providing a service.
- *Requirement (R)* parameters define the properties that a network layer is required to have.

In a layered architecture a layer offers to the layers above services and the quality of these services can be expressed in terms of some parameters of the layer. A parameter can be of type C or type R, depending on whether one takes the perspective of the layer providing the service or that of the layer using it. For example, take the perspective of a layer that requires communication with a maximum delay from a lower layer. In this case the parameter delay is a requirement. From the perspective of the lower layer, delay is a parameter describing its capability. This subtle difference is especially important when the cross-layer optimization is not implemented as a centralized unit, but is performed by a set of local optimizers located on different layers. Matching the requirements and capabilities of layers properly is essential to ensure that the local optimizers cooperate to achieve the global optimization objectives of the network.

6.3 Implementing Cross-Layer Optimization

Cross-layer optimization improves network performance and adaptiveness but may also introduce additional implementation cost compared to a pure layered architecture. There are three main types of cost due to CLD:

- *Computational cost.* CLOs exploring the value assignments of a large set of parameters and evaluating a complex objective function require high computational power and may introduce significant processing delay. Parameter abstraction contributes to handle complexity but may also decrease the optimality of the derived configuration. Another strategy to reduce computational cost is to implement the optimizer as a set of components that operate concurrently and possibly execute on different resources.
- *Communication cost.* CLOs use network parameters that are available at distributed network locations. Gathering these parameters can result in large bandwidth overhead.
- *Reconfiguration and management cost.* Layered architectures are structured as a stack of layers, each defined in isolation and separated from the other layers by well-defined interfaces. Cross-layer architectures are less modular and therefore more difficult to manage or reconfigure if something is to be changed. This type of cost is not easy to quantify; however, it can be limited by also defining interfaces between the traditional layers and the cross-layer optimizers.

Choosing an effective implementation of a cross-layer architecture requires a careful evaluation not only of the performance gain but also of the above cost factors.

The implementation of a CLA can be either centralized or distributed:

- *Centralized.* The CLO is a centralized unit that gathers all the relevant parameters from the network layers, performs the optimization and distributes the selected values of the parameters to the corresponding layers (Figure 6-4). Centralized implementations are often too expensive and inefficient to realize for several reasons. First, gathering network parameters from distributed locations can take time and delay the optimization process. Second, layer parameters vary at rather different rates (of the order of fractions of milliseconds in the physical layer and of the order of seconds in the application layer) and therefore optimizing all the parameters for the worst case may be rather inefficient. Third, computing

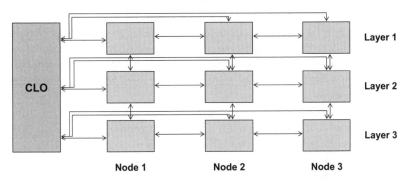

Figure 6-4. Centralized CLO implementation

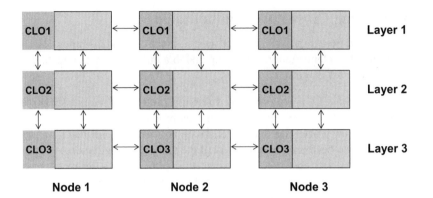

Figure 6-5. Distributed CLO implementation

the objective function for a large number of parameters at the same time may be too expensive.

- *Distributed.* The CLO consists of a set of components distributed across network layers (*vertical distribution*) or nodes (*horizontal distribution*). Each component performs a local optimization over a subset of the parameters of the global optimization problem, but should cooperate with the other components to achieve the global network optimization objective. The implementation shown in Figure 6-5 is based on a distributed optimizer, whose components belong to multiple layers and nodes. Vertically distributed implementations have a hierarchical structure, in which CLO components placed at different layers operate at different rates and optimize local parameters using abstracted representations of the capabilities of the lower layers and of the requirements of the upper layers. As a result, a vertically distributed CLO implementation resembles a layered architecture.

Distributed implementations are usually more practical and efficient, especially when the optimization concerns a large number of parameters. Each CLO component, which is local to a layer or to a node, optimizes a subset of the network parameters. The greatest challenge of designing a distributed implementation is to ensure that all the local optimizers exchange a suited set of parameters through well-defined interfaces and cooperate effectively towards achieving the global optimization objective. In an architecture where the CLO is distributed across multiple layers, each layer includes a local optimizer that selects an assignment of layer parameters while taking into account the requirements from the upper layers and the capabilities of the lower layers. Hence, requirements must be propagated downwards from the application (top-down), while capabilities, expressed in the form of a set of feasible values of parameters, such as error rate, delay and throughput, must be propagated upwards from the lower layers (bottom-up).

Figure 6-6 shows the components of a cross-layer optimization vertically distributed over two layers for channel scheduling in a multi-user video streaming scenario. The optimization in the upper layer is repeatedly performed at the beginning of every Group of Pictures (GOP). The optimizer in the upper layer selects the number of slots that are assigned to each user and the video source rate, based on information provided by the lower layer on the available channel rate during the next GOP period (long-term channel prediction). The selected number

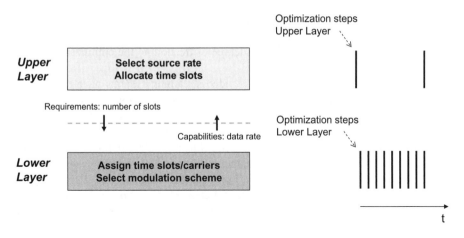

Figure 6-6. Multi-layer optimization: channel scheduling

of slots is then passed as a requirement to the optimizer in the lower layer that assigns specific time slots and carriers and selects the modulation scheme. The optimization in the lower layer is based on short-term channel prediction and therefore is executed at a higher rate.

6.4 Cross-Layer Optimization of Video Streaming Systems

Delivery of high quality video over the wireless medium requires networks that have sufficiently large bandwidth and can adapt to the dynamic variations of the application requirements and of the physical medium capabilities.

Requirements of video streaming applications are rather variable over time. Video data rates vary as a result of the dynamic nature of the captured scene and the encoding policy adopted. A video stream is usually encoded as a sequence of groups of consecutive frames, called Group of Pictures (GOP). In a GOP the first frame is an I-frame and is encoded independently of the other frames. The remaining frames (P-frames, B-frames) are differentially encoded with respect to other frames in the same GOP. A frame can be successfully decoded at the receiver if all the frames in the GOP on which it depends have been correctly received and decoded. The distortion at the receiver varies depending on which frame is lost. For example, the loss of an I-frame results in higher distortion than the loss a P- or a B-frame. Therefore, to optimize wireless video streaming delivery, a network should allocate resources taking into account application parameters such as the relative importance of the packets and the resulting distortion if any packet is lost. At the same time, the application layer should adapt to the time-varying characteristics of the channel, for example by dynamically selecting the rate at the video server that best matches the current transmission capabilities.

Cross-layer optimization allows fast and effective adaptation of a wireless video streaming system. It can be used to optimize individual layers, using knowledge of parameters from other layers. For example, on the basis of predictions of the state of the channels for all the users, together with knowledge of the type of frame carried by each packet at any time, the transmission of the most important frames can be scheduled for transmission over the channels having the best transmission capabilities. Cross-layer optimization is even more effective when used to jointly optimize parameters of multiple layers. For example, selecting the rate of the video stream at the server while allocating the channel resources, instead of just taking the rate as a given parameter, adds another degree of freedom to the optimization. See Khan *et al.*[18] for a comprehensive overview.

Previous applications of CLD to wireless video streaming systems have mostly focused on optimization of individual layer parameters using information from application or physical layers. Krunz and Tripathi[19] allocate channel bandwidth by synchronizing the phase of multiple video streams so that high-peak-rate periods do not overlap in time ('multiplexing gain'). Tupelly *et al.*[15] define an opportunistic scheduling algorithm for multiple video streams, using a priority function that depends on channel conditions, the importance of frames, queue size and multiplexing gain, while Gross *et al.*[13] propose three mechanisms to achieve performance gain in MPEG-4 video delivery over OFDM channels. The first, called semantic queue management, uses knowledge of the relative importance of the packets (I-, P- and B-frames in MPEG-4) to decide which of them can eventually be dropped. The second, called resource allocation, allocates OFDM sub-carriers on the basis of the length of the queues at the base station that stores the packets ready to be transmitted to the mobile terminals. The third assigns sub-carriers to each mobile user. Cell capacity and user-perceived quality are used as performance metrics. Zhang *et al.*[14] present a framework for multimedia delivery over wireless, based on a cross-layer architecture combining application, transport and link layers. The architecture includes a server, a base station and mobile terminals. Functions such as network condition estimation, network-adaptive unequal-error protection, application-adaptive ARQ and priority-based scheduling are discussed for each protocol layer in terms of the parameters they use from other layers. Network-adaptive unequal protection uses information from the application to divide the media into two classes (most important and less important) that are protected from errors in different ways based on the channel conditions.

In the next section we present an example of cross-layer optimization of a wireless video streaming system. The approach jointly optimizes parameters of multiple layers, taking into consideration effective abstractions of the application and physical layers and using an application-based objective function. Furthermore, not only do we present the performance gain of cross-layer optimization through experiments on a testbed, but we also explore the trade-off between the performance gain and the additional computation and communication costs of the optimization.

6.5 Wireless Video Streaming Cross-Layer Architecture

Let us consider an application scenario in which a base station delivers streaming videos to K mobile users located in its cell.[12] We present a CLA where the application, datalink and physical layers are jointly optimized (Figure 6-7). Cross-layer optimization is applied to all the users simultaneously to allocate resources and take advantage of multi-user diversity.[18]

In short, the cross-layer optimization cycle works as follows; details will be provided in the subsequent subsections. First, the CLO takes an abstraction of the parameters of the layers. The physical and data link layers are abstracted on the basis of the transition probabilities of a two-state Gilbert–Elliott model. The application layer is abstracted on the basis of a Rate-Distortion Profile that includes the size of the frames and the expected distortion at the receiver for each type of frame loss.[20] After the process of abstraction, the CLO optimizes

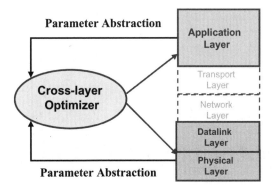

Figure 6-7. Video streaming cross-layer architecture

the system by selecting the optimal values of the video source rate (application layer), the time slot allocation (datalink layer) and the modulation scheme (physical layer).

Peak Signal-to-Noise Ratio (PSNR) is a quantitative parameter that closely represents user-perceived video quality and is therefore selected as a metric for the optimization of video streaming delivery systems. The CLO maximizes the user-perceived video quality measured in terms of the expected PSNR at the receiver. The objective function can be defined in several ways, for example in terms of the PSNR of specific users (perhaps the user in the cell that experiences the worst video quality) or in terms of the average PSNR among all the users in the cell. Once the CLO has selected the optimal values of the parameters, it distributes them to all the individual layers that are responsible for translating them back into actual modes of operation.

6.5.1 Abstracting Layer Parameters

The CLO uses abstractions of the application, the data link and the physical layers.

The application layer is abstracted by the Rate-Distortion profile,[20] which describes the effect of the variations of channel quality on the user-perceived video quality. The Rate-Distortion profile is composed of:

- the Rate Vector, which consists of the size of every frame in the GOP
- The Distortion Matrix, which describes the distortion at the receiver (expressed in Mean Square Error) for each frame loss, assuming that the receiver displays the most recent decoded frame instead of the lost frame (Figure 6-8). The Distortion Matrix is computed at encoding time, stored in the streaming server and sent to the CLO along with the video stream.

Each line is labelled with the most recent decoded frame displayed instead of the lost frame; R indicates the last decoded frame of the previous GOP; the matrix elements show the Distortion D when the frame at the respective position (given as a subscript) is lost and the most recent decoded frame (given as a superscript) is displayed.

The physical layer is abstracted using the Gilbert–Elliot (GE) model, which models the dynamics of the packet-error behaviour of a wireless channel with two states: G (good) and

$$
\begin{array}{c}
R: \\
I: \\
P_1: \\
P_2: \\
B_1: \\
B_3: \\
B_5:
\end{array}
\begin{bmatrix}
D_I^R & D_{B_1}^R & D_{B_2}^R & D_{P_1}^R & D_{B_3}^R & D_{B_4}^R & D_{P_2}^R & D_{B_5}^R & D_{B_6}^R \\
/ & D_{B_1}^I & D_{B_2}^I & D_{P_1}^I & D_{B_3}^I & D_{B_4}^I & D_{P_2}^I & D_{B_5}^I & D_{B_6}^I \\
/ & / & / & / & D_{B_3}^{P_1} & D_{B_4}^{P_1} & D_{P_2}^{P_1} & D_{B_5}^{P_1} & D_{B_6}^{P_1} \\
/ & / & / & / & / & / & / & D_{B_5}^{P_2} & D_{B_6}^{P_2} \\
/ & / & D_{B_2}^{B_1} & / & / & / & / & / & / \\
/ & / & / & / & / & D_{B_4}^{B_3} & / & / & / \\
/ & / & / & / & / & / & / & / & D_{B_6}^{B_5}
\end{bmatrix}
$$

$$...IB_1 B_2 P_1 B_3 B_4 P_2 B_5 B_6 \ IB_1 B_2 P_1 B_3 B_4 P_2 B_5 B_6...$$

GOP i   GOP i+1

Figure 6-8. Rate-distortion (RD) matrix for a sample GOP

B (bad). In the good state, packets are assumed to be received correctly and in a timely manner, while in the bad state packets are assumed to be lost. This model is described by the transition probabilities p from state G to state B and q from state B to G. The transition probabilities (p and q) describing the channel of each user are computed in terms of (1) transmission data rate, (2) transmission packet error rate, (3) data packet size, and (4) channel coherence time, as described by Peng *et al.*[12]

A more abstract representation of the physical layer can be derived from the GE model as follows. Assume that a GOP consists of 15 frames and each frame is transmitted only once. The frame losses can be classified into 16 different patterns. Pattern 1 represents the case where at least one packet in the I-frame is lost and therefore the I-frame cannot be decoded at the receiver. Because of the frame dependencies, none of the frames in the current GOP can be decoded and they will be replaced by the last decoded frame from the previous GOP. Pattern 2 represents the case where all the packets in the I-frame are received correctly but at least one packet in frame P_1 is lost. The other cases can be derived in a similar way. Pattern 16 represents the case without any packet loss. Given the transition probabilities (p and q) of the GE model, the probability of each frame loss pattern p_i is given by

$$
\begin{aligned}
p_1 &= 1 - P_G (1-p)^{(n_1 - 1)}; \\
p_2 &= P_G (1-p)^{(n_1 - 1)} - P_G (1-p)^{(n_1 + n_2 - 1)};
\end{aligned}
$$

$$\cdots$$

$$
p_i = P_G (1-p)^{(n_1 + \ldots + n_{i-1} - 1)} - P_G (1-p)^{(n_1 + \ldots + n_i - 1)}; \tag{6.1}
$$

$$\cdots$$

$$
\begin{aligned}
p_{15} &= P_G (1-p)^{(n_1 + \ldots + n_{14} - 1)} - P_G (1-p)^{(n_1 + \ldots + n_{15} - 1)}; \\
p_{16} &= P_G (1-p)^{(n_1 + \ldots + n_{15} - 1)}
\end{aligned}
$$

where P_G denotes the steady-state probability of being in the good state and n_i ($i = 1, \ldots,$ 15) denotes the number of packets in the ith frame determined from the rate vector and the packet size.[12]

The frame-loss pattern probabilities in equation (6.1) have been derived assuming that each frame is transmitted only once. However, when the transmission rate allocated to a user is larger than the video source rate, the most important frames can be transmitted multiple times to reduce the frame-loss probability. In the case of repeated transmissions, when at least one of the copies of a packet is received correctly, the packet is considered to have been successfully received. When the transmission rate is not sufficient to allow retransmission of all the packets, only the most important ones are retransmitted until the available transmission data rate has been used.

6.5.2 Optimization

At the beginning of each GOP the CLO selects the values of the parameters of the application, datalink and physical layers that maximize the video quality perceived by the users. This requires the CLO to evaluate for each user and for each parameter selection the expected video quality at the receiver in terms of the PSNR, which can be obtained in two alternative ways:

- Computing the expected reconstruction quality at the receiver using the rate-distortion-side information. As the frame-loss pattern probability p_i and the resulting reconstruction distortion D_i are known for each loss pattern, the expected reconstruction distortion D_{exp} is given by $D_{exp} = \sum_{i=1}^{16} p_i D_i$.

- Computing the Expected Number of Decodable Frames (ENDEF) without rate-distortion-side information. This approximation of the expected PSNR is less accurate and therefore results in a less optimal configuration.

Then, once the video quality for each user is derived, the CLO optimizes the network-layer parameters maximizing an objective function that can be defined, for example, in terms of the video quality of the user experiencing the worst video quality among all the users and the average video quality among all the users.

6.5.3 Performance and Cost Analysis

Let us consider a scenario where three users watch the typical test videos, Mother & Daughter (MD), Carphone (CP) and Foreman (FM), that are delivered from a streaming server located at the base station. All three videos are in QCIF resolution (176×144) with a rate of 30 frames/s. The videos are pre-encoded using MPEG-4 at two different target source rates of 100 kbps and 200 kbps. Each GOP has 15 frames, including one I-frame and 14 P-frames. The total transmission capacity of the system is assumed to be 300 k symbols/s. Two different modulation schemes, BPSK and QPSK are considered, for a total rate of 300 kbps and 600 kbps, respectively. Each user has a set of possible transmission rates of {0, 100, 150, 200, 300} kbps and there are 72 possible rate allocations among the three users.

Figure 6-9 compares the performance of the following three cases:

1 Cross-layer optimization is applied and the CLO uses expected PSNR from the Rate Distortion Matrix (RD) (CLO with RD) to evaluate video quality.

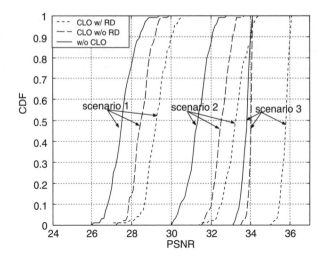

Figure 6-9. CDF of average PSNR

2 Cross-layer optimization is applied and the CLO uses ENDEF (CLO w/o RD) to evaluate
  video quality.
3 No cross-layer optimization is applied (w/o CLO).

The Cumulative Distribution Function (CDF) of the average PSNR is computed for three
scenarios corresponding to different ranges of SNR. In the first scenario all the users have
bad channel conditions (SNR between 0 and 5 dB). Average PSNR is about 2 dB higher
in the case of CLO with RD than in that w/o CLO. In the second scenario, simulations are
performed with random SNR between 0 and 25 dB. In the third scenario all the users have
good channel conditions (SNR between 20 and 25 dB). In all three scenarios we observe an
average PSNR improvement of about 2 dB for cross-layer optimization with RD-side information compared to the case without optimization. The performance of CLO w/o RD-side
information is between the other two cases for all three scenarios. However, in the third scenario the performance of CLO w/o RD is quite similar to that in the case w/o CLO because
of the lower correlation between the number of decodable frames and the resulting
PSNR.[21]

   Our analysis shows that using the expected number of decodable frames still offers a valid
gain with respect to the case w/o CLO, especially in the case of channels with low SNR.
Optimization using the RD profile provides higher performance gain because of the more
accurate calculation of the expected video quality. However, the additional communication
cost due to transmitting the RD profile from the server should also be considered. Figure 6-10
visualizes the communication overhead for different source rates, assuming that the GOP is
composed of an initial I-frame followed only by P-frames. The overhead is quite low, but
increases linearly with the number of frames in a GOP.

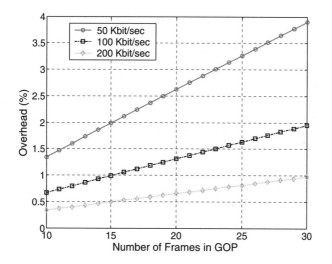

Figure 6-10.   Traffic overhead due to RD profile

6.6 Summary

Cross-Layer Design (CLD) is a new paradigm that has great potential to improve how communication networks will be designed and managed in the future. However, there are several technical challenges that are still to be solved.

So far, CLD has been applied successfully to special cases, which have demonstrated that performance gain can be obtained by optimizing the protocol stack across layers. However, previous work has often neglected the additional cost introduced by cross-layer optimization, which can be especially relevant in resource-constrained systems. The need to evaluate these trade-offs between cost and performance requires the development of system-level methodologies and analysis tools.

Another fundamental question is the implementation of cross-layer optimization. A central optimizer can at best find the optimum settings. However, in real system this is hardly possible to implement, so distributed designs will emerge with local optimizers in each layer that make decisions at a higher frequency compared to a global or central optimizer.

In order to allow a real-time computation of layer settings, an essential aspect is the correct abstraction of parameters describing the settings in the layers. There is a huge trade-off between the optimization quality and the optimization time. A high number of parameters fed into the optimizer will probably lead to a better optimization result, but at the cost of high computation effort. One fundamental research question is to find a low number of abstract parameters that lead to a near-optimum setting.

Choosing the right optimization metric is another essential problem, especially in multimedia applications, where user-perceived quality is subjective and therefore difficult to measure. Furthermore, when multiple applications run concurrently and share network resources, a common objective function must be defined to guarantee the satisfaction of all the users. The Mean Opinion Score (MOS), as used for voice communication, might be a suitable metric, as proposed by Khan *et al.*,[22] but it needs to be mapped to other application classes.

Applying cross-layer optimization usually improves network capacity and increases the number of users being served. However, it is difficult to achieve a full guarantee that QoS constraints are satisfied under all circumstances during network execution in a distributed and highly dynamic environment. Temporary shortages of resources may require the service quality to be reduced for some users or even an interruption in the service. When resources are not sufficient to meet all the constraints, fairness among users should be enforced. Defining fair resource-allocation policies is an active area of research.

Part II

Foundations of Smart Service Provisioning

7

Ontologies

Marko Luther

Recently, attention has been focused on understanding the scientific principles of organizing the vast amount of information stored within data systems. One increasingly popular method for information organization is sought through the construction and implementation of ontologies. An ontology is a formal specification of a conceptualisation,[1] where concepts are distinguished by axioms and definitions describing an area of knowledge.[2] Therefore, an ontology is somewhat similar to a thesaurus, dictionary or glossary, yet with much greater detail and structure, which enables computer to process its content. The word ontology has been used to describe artefact with different degrees of structure. These range from simple taxonomies (such as the Yahoo hierarchy), to metadata schemes (such as the Dublin Core), to logical theories. By establishing a common vocabulary, together with computer-accessible, meaningful descriptions of relevant terms and relationships between these terms, ontologies support the sharing and reuse of formally represented knowledge.

Ontologies are usually expressed in a logic-based language. Consequently detailed, accurate, consistent, sound and meaningful distinctions can be made among the descriptions. In Section 7.1 we introduce some important basic notions of the Description Logics (DLs) formalism that is tailored for representing ontologies together with its associated reasoning services. DLs have a range of applications, but are nowadays mostly known as the formalism that underlies the standard Web Ontology Language (OWL) that figures prominently in the emerging Semantic Web as a way of representing the semantics of documents. OWL is described in Section 7.2. Finally, current methods and tools for the working ontology engineer are sketched in Section 7.3.

7.1 Description Logics

Description Logics (DLs)[3] are a family of general-purpose knowledge representation formalisms based on subsets of first-order predicate logic, where reasoning amounts to the verification of logical consequences. They evolved as a remedy for semantic problems discovered

Towards 4G Technologies. Edited by Hendrik Berndt.
© 2008 John Wiley & Sons, Ltd.

in the ad hoc approaches to knowledge representation, such as semantic network and frame systems, developed in the 1970s.

Research in the area of DLs began under the label *Terminological Systems*, indicating that the representation language was mainly used to establish the basic terminology of a domain. Later, the focus of research shifted towards the concept-forming constructs, giving rise to the name *Concept Logics*. Now, the frameworks are generally known under the term *Description Logics (DLs)*, stressing the importance of the underlying logical system.

Based on a close interaction between theory and practice, research on DLs has successfully covered formal and computational properties of reasoning (Section 7.1.3) as well as implementation of DL-based systems (Section 7.1.4). Those systems have proved useful in many application areas (Section 7.1.5) and have gained a wide acceptance. In particular, the usage of very expressive languages as the formal basis of Semantic Web ontology languages (Section 7.2) has increased acceptance in the last few years.

7.1.1 Basic Description Languages

Most fundamental to DLs is the subsumption relation between concepts that defines the inheritance of properties along with the induced concept hierarchy. A characteristic feature of DLs is their ability to represent other kinds of relationships that can also hold between concepts. However, the more complex concept relationships are, the more difficult it is to characterize precisely the implied consequences, and to completely compute the implicit relationships efficiently.

The DLs approach makes a clear distinction between *intentional knowledge*, expressing general knowledge about a domain that is usually thought not to change, and *extensional knowledge*, which is specific to a particular problem. Intentional knowledge is formulated by terminological axioms that describe general concepts building the so-called T-box, whereas extensional knowledge is formulated by a set of assertional axioms that concern individuals forming the so-called A-box.

Atomic concepts designated by unary predicate symbols, sometimes called *primitive concepts*, and *atomic roles*, designated by binary predicate symbols are the basic ingredients used to define a T-box. Complex terms are then built from the basic symbols using a rather small set of concept- and role-forming constructors. Equalities with atomic concepts on the left-hand side, are used to introduce symbolic names for complex descriptions and are therefore called definitions:

$$\text{Person} \equiv \text{Man} \sqcup \text{Woman}$$
$$\text{Mother} \equiv \text{Woman} \sqcap \exists \, child \, . \, \text{Person}$$

While definitions such as the one above define necessary and sufficient conditions of a concept, primitive definitions stating inclusion can be used to describe only necessary ones:

$$\text{Woman} \sqsubseteq \text{Person}$$

Axioms of the form $C \sqsubseteq D$ for a complex description C are often called general concept inclusion axioms (GCI).

Description Logics can be differentiated by the set of role- and concept-forming operations they provide. The *Attributive Language* $(\mathcal{AL})$[4] comes with atomic concepts (A), a universal concept ($\top$), a bottom concept ($\bot$), negation that is only applicable to atomic concepts ($\neg A$), intersection (C $\sqcap$ D), universal value restriction ($\forall R.C$) and limited existential quantification ($\exists R.\top$). Another prominent DL is the *Simple Propositional Description Logics* ($\mathcal{ALC}$) that corresponds to the fragment of first-order logic obtained by restricting the syntax to formulas containing two variables.[3] $\mathcal{ALC}$ extends $\mathcal{AL}$ by full complement (i.e. concept negation) and allows for disjunction (C $\sqcup$ D), and full existential quantification ($\exists R.C$). The *Frame Language* ($\mathcal{FL}^-$), also called *Structural Description Logic*, is obtained from $\mathcal{AL}$ by disallowing atomic negation. As a minimal language of practical interest, $\mathcal{FL}_0$ is obtained from $\mathcal{FL}^-$ by disallowing limited existential quantification.

Generally, DL languages are categorized by constructing mnemonic names that encode the precise expressivity of the logic. Adding corresponding letters to the name of the system indicates that a DL language is extended by additional constructors. In order to avoid very long names for expressive DLs, the abbreviation S was introduced for $\mathcal{ALCR}^+$, that is, the DL that extends $\mathcal{ALC}$ by including transitively closed primitive,[5] as indicated by adding the suffix $\mathcal{R}^+$ (see Table 7-1).

An A-box contains extensional knowledge consisting of individual axioms sometimes called facts. For example:

$$alice \in Woman$$

states that alice is a Woman. Similarly,

$$alice \; child \; bob$$

specifies that bob is a child of alice. Assertions of the first kind are called *concept assertions*, while assertions of the second kind are called *role assertions*. Inference services on A- and T-boxes, as the one that computes the taxonomy (note that some authors use the name taxonomy as a synonym for ontology) of the terminological and assertional axioms, given in this section and depicted in Table 7-1, are introduced in the next section.

Table 7-1. The description logic S

Construct name	Syntax	Example
Atomic concept	A	Person
Universal concept	$\top$	$\top$
Subsumption	(C $\sqsubseteq$ D)	Man $\sqsubseteq$ Person
Equivalence	(C $\equiv$ D)	Child $\equiv$ Kid
Atomic role	R	Child
Transitive role	*Trans(R)*	friend
Conjunction	C $\sqcap$ D	Person $\sqcap$ Female
Disjunction ($\mathcal{U}$)	C $\sqcup$ D	Man $\sqcup$ Woman
Negation (C)	$\neg$ C	$\neg$ Male
Existential qualification ($\mathcal{E}$)	$\exists R.C$	$\exists$ child.Person
Value restriction	$\forall R.C$	$\forall$ child.Female

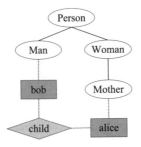

Figure 7-1. Taxonomy

7.1.2 Inference Services

Ontology-based reasoning is concerned with drawing conclusions from formal concept, role and individual descriptions. Implicit knowledge about concepts and individuals can be inferred automatically with the help of sound and complete inference algorithms, which are known for a wide variety of DLs (see Section 7.1.3). In particular, relationships between concepts, as well as instance relationships between individuals and concepts, play an important role.

7.1.2.1 Standard Inferences

The basic inference on concept expressions in DLs is *subsumption*, expressing that the denotation of one concept is a subset of the other. Another typical reasoning task, important at design time, is to determine whether a description is *satisfiable* (i.e. non-contradictory). A *classification* process organizes the concepts of a terminology into a lattice-like structure defined by the subsumption relationship. A-box reasoning is about finding out whether its set of assertions is *consistent* (i.e. whether it has a model) and whether a particular individual is an instance of a given concept description. The basic inference tasks described in this paragraph are listed in Table 7-2, together with other typical queries that can be posed against a knowledge base. However, for practical reasons, many DL systems support additional queries such as the retrieval of the set of individuals or concept names (see Section 7.1.4).

During the development phase, queries about the logical consistency are especially important to ensure the quality of the resulting knowledge base. Pure modelling can result in inconsistent concept definitions in the T-box, describing an empty set of individuals. Such inconsistent definitions cause inconsistent knowledge bases that affect every sentence, making ontology reasoning useless. Likewise, an inconsistency can also occur in the A-Box of a knowledge base, if given individual restrictions are contradictory with respect to the T-box.

Note that both the consistency check and the classification process always concern the whole T-box, which can lead to drastic impacts on a knowledge base; even though a knowledge base can be considered consistent – for instance after a phase of knowledge acquisition in which only sound concepts have been added – re-classification can still result in a significant restructuring of the existing T-box.

Table 7-2. Standard inference services

Reasoning task	Description	Effects
Concept consistency	Is the set of objects described by a concept empty?	T-Box
Concept subsumption	Is there a subset relationship between the set of objects described by two concepts?	
Consistency check	Find all inconsistent concepts mentioned in a T-box. Inconsistent concepts may be the result of poor modelling or data gathering errors.	
Classification	Determine the parents and children of a concept with respect to a T-box. The parents of a concept are the most specific concept names mentioned in a T-box that subsume the concept. The children of a concept are the most general concept names mentioned in a T-box that the concept subsumes.	
A-box consistency	Are the restrictions given in an A-box with respect to a T-box too strong, i.e. do they contradict each other?	A-Box
Instance testing	Is the object for which an individual stands a member of the set of objects described by a certain query concept? The individual is then called an instance of the query concept.	
Instance retrieval	Find all individuals from an A-box such that the objects they stand for can be proven to be a member of a set of objects described by a certain query concept.	
Individual direct types	Find the most specific concept names from a T-box of which a given individual is an instance.	

7.1.2.2 Open vs Closed-World Assumption

A distinguishing feature of DLs is that they make the Open-World Assumption (OWA). While a database instance represents exactly one interpretation, namely the one where classes and relations in the schema are interpreted by the objects and tuples in the instance, an A-box represents many different interpretations, namely all its models. Therefore, absence of information in a database instance is interpreted as negative information, while absence of information in an A-box only indicates a lack of knowledge. Consequently, the information in a database is always understood to be complete, whereas the information in an A-box is in general considered as incomplete. The view that A-boxes have an open-world semantics is clearly reflected in the DL technology, which employs the OWA for reasoning. This means that 'what cannot be proven to be true is not believed to be false'. Therefore, a negative answer stands for 'it cannot be proven with respect to the given information'.

As depicted in Figure 7-2, the absence of information regarding another person in the vicinity of miller is not interpreted as 'there is none'. Even though all known individuals associated with the role 'together' are instances of the concept 'Colleague', it cannot be concluded under the OWA that the individual 'miller' fulfils the specification of the concept 'AtWork'. The OWA ensures monotonicity of the entailment relation with respect to additional assertions, such as 'miller together wallace' and 'wallace $\sqsubseteq \neg$ Colleague', which might be added to the A-box later on.

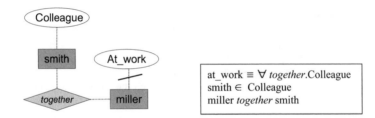

Figure 7-2. Open-World Assumption

Table 7-3. Extensions of S

Construct Name	Syntax	Example	System
Role hierarchy	$R \sqsubseteq S$	father $\sqsubseteq$ parent	$\mathcal{H}$
Inverse role	R^-	supervises = supervisor$^-$	$\mathcal{I}$
Functional role	$Funct(R)$	father	$\mathcal{F}$
Unqualified number restriction	$\geq n\ R$	≥ 3 child	$\mathcal{N}$
	$\leq n\ R$		
Qualifying number restriction	$\geq n\ R . C$	≥ 3 child . Female	$\mathcal{Q}$
	$\leq n\ R . C$		
Nominal	$I_1 \sqcup \ldots \sqcup I_n$	{red,green,blue}	$\mathcal{O}$
Concrete domains	$u_1, \ldots, u_n . P$	temp. (>20)	$(\mathcal{D})$

In contrast, the Closed-World Assumption (CWA) states that everything that is not known or cannot be proved to be true is assumed to be false. CWA originates from AI and database research in the late 1970s and the same fundamental assumption still holds for programming languages like Prolog and the design most of databases today. The assumption is useful in that it allows additional inferences to be drawn from the absence of information. However, an assumption that any unknown facts must be false may sometimes be mistaken.

7.1.3 Language Extensions

Separating out the structure of concepts and roles into simple term-forming operators opened the door to extensive analysis of a broad family of languages. As a result, the family of DLs is probably the most thoroughly understood set of knowledge representation formalisms. The computational space has been thoroughly analysed, triggered by the trade-off between the expressiveness of the representation language and the complexity of reasoning.

In the previous section we introduced the basic DL concept and role constructors. However, most DLs in current use provide further language features, which are often indicated by adding additional letters to the system's name (see Table 7-3). The letter $\mathcal{H}$ is used for role hierarchies with multiple parents. If a language supports inverse roles, this is indicated with the letter $\mathcal{I}$, and, if it allows for functional restricted properties, the letter $\mathcal{F}$ is added. $\mathcal{N}$ stands for simple number restrictions and the letter $\mathcal{Q}$ is used for qualified number restrictions. The letter $\mathcal{O}$ denotes language constructors for an extensional specification of concepts using nominals.

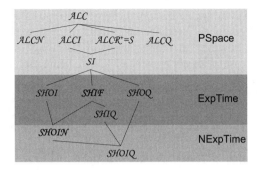

Figure 7-3. Important description logics

To overcome the limited support of most DLs for specific domains such as concrete data types (such as numbers or strings), spatial regions or qualitative time intervals, Baader and Hanschke defined a general method for integrating concrete domains into DLs indicated by adding ($\mathcal{D}$) to the system's name.[6]

Prominent members of the S-family are $\mathcal{SH}$ (which extends $\mathcal{ALCR^+}$ with role hierarchies),[7] $\mathcal{SHIF}$ (which extends $\mathcal{ALCR^+}$ with number restrictions, inverse roles and number restrictions of the form (≤ 1 R))[8] and $\mathcal{SHIQ}$ (which is pronounced $\mathcal{ALC}$-choir and extends $\mathcal{ALCR^+}$ with role hierarchies, inverse roles and qualified number restrictions); see Figure 7-3.[9]

While reasoning for S extended by inverse roles is still of worst-case polynomial space (PSpace),[5] this upper bound is not robust w.r.t. extensions (see Table 7-3). Reasoning is already worst-case deterministic exponential time (ExpTime) for $\mathcal{SHOQ}(\mathcal{D})$,[10] $\mathcal{SHOI}$[11] and even for systems without nominals such as $\mathcal{SHIF}$ and $\mathcal{SHIQ}$.[9] For systems combining nominals with concrete domains such as $\mathcal{SHOIN}(\mathcal{D})$ (i.e. $\mathcal{SHIQ}$ restricted to unqualified number restrictions and extended by nominals and conrete domains),[12] the complexity of reasoning is even higher, that is nondeterministic exponential time in the worst-case (NExpTime).

Although adding a feature does not change the worst-case complexity of a system, it may influence practical implementations. For example, inverse properties in very strong DLs make several important optimization techniques much less effective.[13] The difficulty in further extending the $\mathcal{SHOQ}$ or $\mathcal{SHIQ}$ optimization algorithms to $\mathcal{SHOIQ}$ (i.e. $\mathcal{SHOIN}$ extended with qualifying number restrictions) due to the interaction between nominals, number restrictions and inverse roles,[14] has only been overcome very recently.[15]

7.1.4 Description Logics Systems

The ancestor of all DL systems is KL-ONE,[16] which initiated the transition from semantic networks to well-founded DLs. As the more expressive successors such as LOOM,[17] subsumption in KL-ONE was decided structurally. However, while they were sound, the structural algorithms were found to be incomplete in terms of the logical semantics. Later, the need for complete inference services was the focus of attention, leading to the system KRIS,[18] which introduced optimized classification algorithms based on caching and propagation techniques. More recently, the specialization of classical tableau calculi led directly to practical implementations of complete inference algorithms for very expressive, but intractable, DLs.[4,7] It has turned out, however, that intractability of reasoning (in the sense of being non-polynomial in the worst case) does not prevent a DL from being useful in practice. It provides sophisticated optimization techniques used in implementations, such as lazy unfolding, absorption and dependency directed backtracking.[7] Current complete and efficient DL systems for very expressive DLs are FaCT,[19] RACER[20] and Pellet.[21] These systems offer more or less all of the inference services discussed in Section 7.1.2.1 and are introduced here in greater detail.

FaCT (Fast Classification of Terminologies) is a highly optimized DL classifier written in Common Lisp for terminologies expressed in the $\mathcal{SHIQ}$ logic.[19] It supports reasoning with knowledge bases containing CGIs, but cannot deal with individuals or concrete datatype domains. Furthermore, FaCT does not support multiple T-boxes and does not provide mechanisms for removing concept definitions. With FaCT++ a reimplementation of FaCT in C++ is available.[22] It implements the logic $\mathcal{SHIF}(\mathcal{D}-)$ and is based on a new internal architecture that introduces new optimizations. While FaCT++ implements A-box reasoning, the support for concrete domains is restricted to Integers and Strings (therefore $\mathcal{D}-$).

RACER is another highly optimized tableau calculus reasoner implemented in Common Lisp.[20] It supports all optimization techniques that are incorporated into FaCT, plus some additional optimizations dealing with number restrictions and A-box reasoning, which are dynamically selected based on a static analysis of the given knowledge base and query. Like FaCT it implements $\mathcal{SHIQ}(\mathcal{D}-)$, but additionally supports A-box reasoning services and retraction of A-Box assertions, as well as multiple T- and A-boxes. Furthermore, RACER provides facilities for algebraic reasoning, including concrete domains for dealing with restrictions over integers, linear polynomial equations over reals, nonlinear multivariate polynomial equations over complex numbers and equalities and inequalities of strings. However, Racer does not support complete reasoning with respect to nominals. Transforming individuals in the enumerated classes to disjoint atomic concepts, as is done by RACER, only approximates nominals.

Pellet is a sound and complete tableau reasoner for $\mathcal{SHIN}(\mathcal{D})$ as well as $\mathcal{SHON}(\mathcal{D})$ and a sound but incomplete reasoner for $\mathcal{SHION}(\mathcal{D})$.[21] Additionally, Pellet provides sound and complete reasoning for nominals by using the algorithms developed for $\mathcal{SHOQ}(\mathcal{D})$.[12] It is implemented in Java and incorporates a number of special features. Pellet supports ontology analysis and repair, datatype reasoning for built-in primitive XML Schema datatypes, conjunctive A-box query and a direct implementation of entailment checking.

7.1.5 Applications

DLs have been used in a range of applications and concrete systems.[3,23] One of the first application domains for DLs has been in the field of Software Engineering, undertaken at AT&T to implement a Software Information System specifying facts about large software systems. Other very successful applications in that domain support developers in the design of complex systems by verifying certain properties on component configurations.

Medicine is the domain where Expert Systems have been developed since the early 1980's. The ability to represent and reason about taxonomies in DLs has motivated the construction and maintenance of very large ontologies of medical knowledge. Besides further application areas such as natural language processing and database management, there have been significant efforts based on the use of markup languages to semantically annotate the information content of Web structures.[24] The use of DLs in the design of Semantic Web applications is addressed in the next chapter.

7.2 Web Ontology Language

It has been predicted that ontologies will play a pivotal role in the Semantic Web. The World Wide Web Consortium (W3C) sketched the vision of a 'second generation' Web in which Web resources will be more readily accessible to automated processes.[25] A key component of the Semantic Web will be the annotation of web resources with machine-accessible metadata that describes their content, with ontologies providing a source of shared and precisely defined terms that can be used in such metadata. This requirement has led to the extension of Web markup languages in order to facilitate content description and the development of Web-based ontologies such as XML Schema, RDF (Resource Description Framework) and RDF Schema (RDFS).

RDF is a general-purpose language for annotating resources on the Web, while RDFS is a schema language that allows the definition of properties and kinds of the resources. RDFS in particular is recognizable as an ontology representation language; it allows the organization of vocabulary in typed hierarchies by defining classes and properties (binary ground relations), range and domain constraints (on properties), instances of classes, and subclass and subproperty relationships. However, the expressive power of RDFS is rather limited and the reasoning support is limited to constraint checking. It misses constructs to define class disjointness, Boolean combinations of classes, special characteristics of properties (such as transitivity) and lacks the possibility of restricting the number of distinct values a property may or must take (i.e. cardinality restrictions).

To overcome the limited expressiveness of RDFS, a more expressive Web Ontology Language (OWL)[26] has been defined by the W3C based on the former proposal DAML + OIL,[27] which itself is the result of merging the frame-based American proposal DAML-ONT with the DL-based European language OIL (see Figure 7-4). To fulfil the aim for a standardized

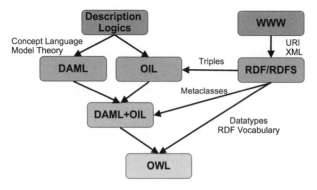

Figure 7-4. OWL influences

and broadly accepted ontology language for the Semantic Web, the language needed a well-defined syntax and semantics with sufficient expressive power and efficient reasoning support. Formal semantics and reasoning support is usually provided by mapping the ontology language to a known logical formalism. In the case of OWL, expressive Description Logics (DLs), as introduced in the previous section, have been selected as the semantic foundation, enabling efficient reasoning by the application of existing DL reasoners.

OWL is layered on top of XML, RDF and RDFS.[28] Syntactically, OWL can be seen as a specific dialect of RDFS (for example, OWL individuals are defined using RDF descriptions). Each valid OWL document has to be a valid RDFS document (but not vice versa). However, this XML/RDF syntax of OWL does not provide a very readable syntax, as shown in the examples given in the next section. Therefore, the more readable OWL abstract syntax has been defined.[29]

Ideally, the OWL semantic would be an extension of that of RDFS, in the sense that OWL would use the RDF meaning of classes and properties and would add language primitives to support the richer expressiveness identified above. However, these additional primitives stemming from DLs lead to uncontrollable computational properties if combined with the higher-order features of RDFS. Therefore, OWL is defined as three different sublanguages (see Section 7.2.1), each fulfilling different aspects of the incompatible requirements. OWL Lite and OWL DL are basically very expressive Description Logics with an RDF syntax. OWL Full incorporates maximum expressive power and syntactic freedom by covering all of RDFS, but has serious logical problems. The lack of a semantic distinction between classes and individuals in the RDFS part of OWL Full leads to the undecidablility of key inference problems (for example, Russell's Paradox is expressible).

7.2.1 Language Elements

While OWL ontologies are basically RDF documents, the *header* of an OWL document is an `rdf:RDF` element, which specifies a number of namespaces. The ontology itself starts with a collection of assertions grouped under an `owl:Ontology` element, which contains comments, version control and the inclusion of other Ontologies by `owl:import` statements. The import statement is transitive with a textual inclusion semantic. The following fragment is a valid OWL header of the ontology `humans.owl` that imports the ontology `gender.owl`:

```
<?xml version='1.0'?>
<rdf:RDF
    xmlns:rdfs='http://www.w3.org/2000/01/rdf-schema#'
    xmlns:owl='http://www.w3.org/2002/07/owl#'
    xmlns:agent='http://localhost/Ontologies/0.1/agent.owl#'
    xmlns:rdf='http://www.w3.org/1999/02/22-rdf-syntax-ns#'
    xmlns:xsd='http://www.w3.org/2001/XMLSchema#'
    xmlns='http://localhost/humans.owl#'
  xml:base='http://localhost/humans.owl'>
  <owl:Ontology rdf:about=''>
    <owl:imports rdf:resource='http://localhost/gender.owl'/>
  </owl:Ontology>
```

Classes are defined using an `owl:Class` element, essentially a subclass of `rdfs:Class`. The following fragment defines `Woman` as a subclass of `Human` disjoint from the class `Man` (by using the class constructors `owl:subclassOf` and `owl:disjointWith`):

```
<owl:Class rdf:about='#Woman'>
  <rdfs:subClassOf rdf:resource='#Human'/>
  <owl:disjointWith rdf:resource='#Man'/>
</owl:Class>
```

Equivalences between classes are expressed by using an `owl:equivalentClass` element. Furthermore, it is possible to define Boolean combinations of classes using the class forming constructs `owl:complementOf`, `owl:disjointWith` and `owl:intersectionOf`. Finally, there are two predefined classes, `owl:Thing` and `owl:Nothing` corresponding to the top and bottom elements in Description Logics (see Section 7.1.1).

Properties can be used to pose certain restrictions on a class definition using the constructs `owl:Restriction` and `owl:onProperty`. For example, the following OWL fragment defines the concept `Mother` as a woman who has some children (i.e. the intersection of the class `Woman` with the restriction that all members need to be related with some individuals of the concept `Human` by the property `has_child`).

```
<owl:Class rdf:ID='Mother'>
    <owl:equivalentClass>
      <owl:Class>
        <owl:intersectionOf rdf:parseType='Collection'>
          <owl:Class rdf:ID='Woman'/>
          <owl:Restriction>
            <owl:someValuesFrom rdf:resource='#Human'/>
            <owl:onProperty>
              <owl:ObjectProperty rdf:ID='child'/>
            </owl:onProperty>
          </owl:Restriction>
        </owl:intersectionOf>
      </owl:Class>
    </owl:equivalentClass>
  </owl:Class>
```

The OWL class restriction declaration `owl:someValuesFrom` corresponds to an existential quantification and the corresponding `owl:allValuesFrom` to a universal quantification. Other types of restriction declarations available are the cardinality restrictions `owl:minCardinality`, `owl:maxCardinality` and `owl:cardinality`. In OWL there are two kinds of properties: Object properties bind objects to objects as the property `child` in the example above, while datatype properties relate objects to datatype values. Properties are defined using the constructs `owl:ObjectProperty` and `owl:DatatypeProperty` and may contain range and domain restrictions (`rdfs:range` and `rdfs:domain`). Furthermore, using the OWL elements `inverseOf`, `subPropertyOf`, `TransitiveProperty`, `SymmetricProperty`, `FunctionalProperty` and `equivalentProperty`, a property can be related to its inverse. A subproperty can be described as symmetric, functional or transitive and, additionally, equivalences of properties can be

defined. The following fragment defines the object property `child` as a subproperty of the
property `relative`, which relates humans and has the property `parent` as inverse:

```
<owl:ObjectProperty rdf:about='#child'>
    <rdfs:range rdf:resource='#Human'/>
    <rdfs:domain rdf:resource='#Human'/>
    <owl:inverseOf rdf:resource='#parent'/>
    <rdfs:subPropertyOf rdf:resource='#relative'/>
</owl:ObjectProperty>
```

Enumerations, also called nominals or value sets, are defined using the constructor `owl:`
`oneOf` constructor. It allows a class to be specified by listing all its elements as depicted in
the following example:

```
<owl:Class rdf:ID='Gender'>
  <owl:one of rdf:parseType='Collection'>
    <owl:Thing rdf:about='#male'/>
    <owl:Thing rdf:about='#female/>
  </owl:oneOf>
</owl:Class>
```

Individuals of classes, sometimes called instances, are declared as in RDF. For example, the
following fragment specifies the individual `alice` to be a woman who has a son called
bob:

```
<Woman rdf:ID='alice'>
    <has_child>
        <Man rdf:ID='bob'/>
    </has_child>
</Woman>
```

The OWL language reference document[26] contains further details on the language constructs
introduced here and describes other constructs that make it possible to describe datatypes,
annotations, the disjointness of individuals and versioning information, among other features.
Note that OWL does not make the Unique Name Assumption (UNA). In contrast, it is
assumed that two object names denote different things under DL semantics.

7.2.2 Sublanguages

OWL is declined into three increasingly expressive sublanguages (see Figure 7-5), which are
respectively OWL Lite, OWL DL and OWL Full.[30] OWL Full is the most expressive language,
which allows all of the OWL constructs discussed in the previous section and covers all of
RDFS. Because of the lack of a clear separation between classes and individuals in its sub-
language RDFS, OWL Full comprises some higher-order features that result in the undecidability of key inference problems.[9] OWL DL, a decidable fragment of first-order logic, uses
exactly the same vocabulary as OWL Full, but includes some additional syntactic restrictions.
With these restrictions, OWL DL corresponds to the Description Logic (DL) $\mathcal{SHOIN}(\mathcal{D})$,
decidable in NExpTime (see Section 7.1.3). OWL Lite is a fraction of OWL DL aiming to

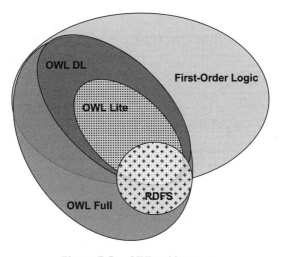

Figure 7-5. OWL sublanguages

provide a useful subset of language features that are easier to present to naïve users. The expressivity of OWL Lite is about that of the DL language $\mathcal{SHIF}(\mathcal{D})$ with an ExpTime worst-case complexity (cf. Section 7.1.3). In contrast to OWL DL, OWL Lite excludes syntax constructs for unions, complements, and individuals in descriptions or class axioms. In addition, it limits cardinalities to either 0 or 1 and nested descriptions to concept identifiers. However, these restrictions come with relatively little loss in expressive power. With help of indirection and syntactical tricks, all of OWL DL can be captured in OWL Lite, except those descriptions containing either individuals or cardinalities greater than 1.[28]

Since nominals are non-standard with respect to traditional DL research, practical complete algorithms for the full OWL DL language were unknown for a long time before the recent result of Horrocks and Sattler.[15] However, this work still leaves the question open how to implement a sound, complete and high-performance reasoning system supporting the XML Schema Datatypes of OWL that goes beyond the standard concrete domains (Integer, Real, String) of traditional DL languages. The logic containments are shown in Figure 7-6.

7.2.3 Rule-based Extensions

Because of its focus on decidability, there are a number of things that cannot be represented in OWL DL. Many of these involve property chaining, i.e. the ability to express constraints among multiple properties (for example the uncle property as the composition of parent and brother). In addition to DL-based reasoning in the T-box and A-box of an ontology, it can be desirable to interface with ontologies at a higher level to overcome some of the expressive restrictions of OWL. To this end, query and rule languages for ontologies are developed, where the uncle property could be expressed as follows, using variables to match individuals of the A-box:

$$\text{parent}(?x, ?y) \text{ and brother}(?y, ?z) => \text{uncle}(?x, ?z)$$

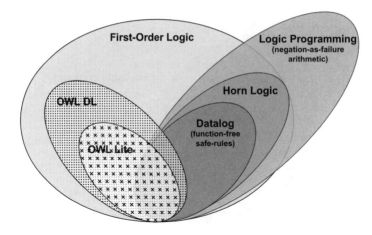

Figure 7-6. Logic containments

However, it is important to stress that, given the expressivity of DLs, query answering cannot simply be reduced to model checking as in the database framework. In fact, query answering in the DL setting requires the same reasoning machinery as logical derivation.[31] Current systems based on DL usually offer only a weak query language that is restricted to retrieval (look up the instances of a given concept), realization (determine the most specific concept an individual is an instance of) and instantiation (Boolean query asking if a pair of individuals is an instance of a given role). In particular, the use of variables to formulate a query is not supported. Recent results show that a conjunctive query language can enrich DL systems with certain restrictions regarding the use of variables in the query, providing a solution for one weakness of traditional DL systems. Completely removing these restrictions causes problems, particularly if variables are used to force cycles in the query.[31] Several proposals to complement ontology reasoning by querying languages already exist. DQL[32] is a formal language supporting query-answering in DAML + OIL ontologies, whereas OWL-QL[33] is an adaptation of DQL that works with OWL. The reasoner RACER has recently been extended with its own new RACER Query Language nRQL,[34] which supports the retrieval of variable bindings in arbitrary concept and role expressions.

Declarative languages for querying and specifying views over RDF/S description bases have also been proposed in the literature; these include RQL,[35] RDQL[36] and TRIPLE[37] (see Haase *et al.*[38] for an extensive comparison). However, unlike those languages, ontology-based query languages support query-answering dialogues in which automated reasoning methods may be used to derive answers.

Rule systems consist of a rule base (i.e. an unordered set of if–then rules), a working memory (i.e. a set of facts represented as literals) and a rule engine that implements forward or backward chaining. The 'if' portion of a rule is a series of patterns that specifies the facts (or data) causing the rule to be applicable. The process of matching facts to patterns is called pattern matching. Rule systems used as a knowledge representation format have the advantage of simplicity and modularity. Because rules are considered to be independent of each other, at least to a certain degree, rule bases are easy to maintain and update. Furthermore, generic rule reasoning with a most-specific-first conflict resolution strategy enables simple default reasoning. However, default reasoning is inherently non-monotonic and undecidable (see the standard work of Reiter, McCarthy and McDermott on non-monotonic reasoning). Additionally, most rule systems lack a clearly defined semantics. Therefore, assuring the reliability of these systems by showing properties such as consistency of the rule base or termination and confluence of the derivation process can be a challenging task.[39] A (static) analysis has to cover the whole reasoning system, because the realized consequence relation depends on both the rule engine and the concrete rule base. In particular, rule interactions (for example two applicable rules with inconsistent conclusions) are hard to avoid and can be troublesome. They make the meaning of a rule set sensitive to the conflict resolution strategy. As Patrik Winston has put it 'the advantage of bequeathing control becomes the disadvantage of losing control'.[40]

Realizing OWL-reasoning by using a specialized rule set is computationally very expensive and reasoning about concepts has to be done indirectly by creating prototypical instances.[41] Furthermore, while being sound, this rule-based approach cannot be complete (not even for OWL Lite[42]). The Jena manual states that this approach is only suited to applications primarily involving instance reasoning with lightweight, regular ontologies.[41]

Clearly, a combination of DL and rule reasoning is desirable, because a generalized form of rules containing complex DL expressions is significantly more expressive than Horn rules.[17] In particular, these generalized rules can express existential and negation information that is not expressible within the (decidable) Horn logic subclass of first-order logic (see Figure 7-7). However, such a tight combination might easily lead to the undecidability of interesting reasoning problems, as rules can be used to simulate role value maps.[43] A basic example that leads to non-termination is given by Motik et al.[44] Possible interactions between description logics and monotonic rule systems were studied by Grosof et al.[45] Based on that work and on the previous work on hybrid reasoning,[17,46] it seems that one option is to take the intersection of the expressive power of DL and function-free Horn rules. However, the straightforward combination of DL with Horn rules, as in the proposal for a Semantic Web rule language SWRL,[47] can be used to simulate Turing machines,[46] which clearly leads to undecidability. This restriction on expressivity is formulated in Description Logic Programs,[43] which prohibits existentially quantified knowledge in consequents. Decidability can also be

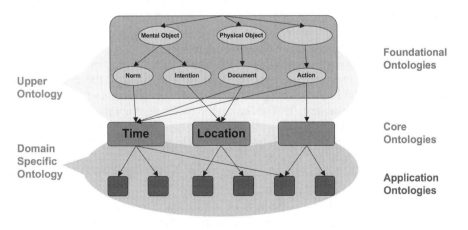

Figure 7-7. Ontology layering

retained by restricting rules to so-called DL-safe ones, requiring each variable in a rule to occur in a non-DL-atom in the rule body.[44]

Alternatively, DL reasoning can loosely be coupled with rule reasoning by deriving DL consequences first and applying rules on the results in a second step. This way, however, some conclusions are not derivable, even if they are semantically entailed. For example, this method cannot derive the fact 'Child(abel)' from the A-box assertion 'abel isA (GoodChild or BadChild)' and the two rules 'if GoodChild(x) then Child(x)' and 'if BadChild(x) then Child(x)'.

Promising rule extensions are the aforementioned OWL rule language SWRL and Racer's built-in query language nRQL. SWRL allows the formulation of event–condition–action rules to automatically trigger actions in applications based on the derived conclusions, but is in general undecidable. In contrast, nRQL lacks the possibility to trigger actions directly, but is decidable and tightly integrated in the core DL-reasoning engine. However, nRQL provides a publish-and-subscribe mechanism that allows clients to be informed about changes in the knowledge base.

7.2.4 Language Deficiencies

The functionality of OWL's `owl:import` construct is unsatisfactory for modularization purposes. Its use essentially results in a completely flat ontology by merging all the linked ontologies into a single logical space.[48] Furthermore, the import of ontologies is not species safe in OWL. For example, importing one OWL DL document into another may result in an OWL Full document. One solution may be an extension of OWL that enables distributed tableaux reasoning on interconnected Ontologies.[49]

Another issue is related to OWL's Open World Assumption (OWA).[3,40] In the W3C OWL requirements recommendation,[50] the ability to state closed worlds is seen as an objective, rather than a requirement. The advantage of the OWA is the monotonicity of the reasoning. Previous reasoning results still hold if additional information becomes available. The disadvantage is, however, that some conclusions cannot be derived under the conservative OWA. A promising approach to integrate closed-world aspects into open-world systems emerged in the agents' community as the Local Closed-World Assumption (LCW).[51] With LCW, closed-world information can be obtained on subsets of the information that is known to be complete, while still allowing other information to be treated as unknown. However, it is still unclear how LCW can be made compatible with the DL approach of OWL.

The currently available reasoning support also suffers from a few drawbacks. Most of the available reasoners do not support OWL completely. For example, the popular RACER system lacks full support for the owl:import statement. This implies that some interconnected ontologies cannot be loaded directly. Furthermore, RACER does not implement incremental reasoning,[52] making a complete reclassification necessary after each context change. Similarly, the standard Description Logic Implementation Group's DIG interface,[53] implemented by most DL reasoning tools and editors, does not support the removal of specific parts of an ontology, a case occurring rather often in dynamic applications, making it necessary to reload (and reclassify) all ontologies if properties are modified.

Finally, adding complex class definitions or storing and querying over large numbers of individuals, common in many application areas, degrades the performance of existing reasoners to the point where they are no longer applicable. There are some investigations that are trying to solve this problem, such as the Instance Store[11] and LAS[54] making use of a hyprid DL-DB architecture. However, those solutions have some restrictions. For example, the Instance Store is only able to deal with data that does not include any role assertions.

7.3 Ontology Engineering

Building an ontology for a fairly large domain can easily become a huge and complex task. Such big undertakings cannot be a success without the existence of a proper methodology, which guides the development as with any software engineering artefact. Ontological Engineering refers to the set of activities that concern the ontology development process, the ontology life cycle, the methods and methodologies for building ontologies, and the tool suites and languages that support them.

7.3.1 Design Principles

The following design guidelines, formulated by Gruber,[55] address issues that have been identified as the most relevant for any ontology development process:

- *Scope.* The ontology development should focus on the selected domain.
- *Clarity.* The intended meaning of terms should be communicated effectively. The ontologies should, therefore, be simple enough to be easily used by application developers and they should enable practical, meaningful, intuitive and simple queries. A detailed ontology, which is completely expressive, tends not to be very useful if it is too complex compared to the necessary level of detail required by most applications.
- *Coherence.* The definitions given should be consistent. Not only should they be logically consistent, but the informal part of the ontology should also be consistent with the formal part.
- *Extendibility.* The ontology should be designed with extendibility in mind. It should be easy to add new terms without having to revise the existing definitions. This enables ontologies in specialized domains to build upon more general ontologies that have already been defined.
- *Minimal encoding bias.* Representation choices should not be made for the convenience of notation or implementation. Ontologies should be as independent as possible from the applications that will use the ontology and of the language in which they are formulated.
- *Minimal ontological commitment.* To make an ontology as reusable as possible, the ontology should make as few assumptions as possible about the world, while still supporting the intended use.

7.3.1.1 Language Selection

The language used for specifying an ontology restricts the expressiveness of the ontology as well as its usability and reusability across applications. For example, a language such as OWL Full is generally undecidable and hard to handle. The consistency of the representation cannot be derived automatically and one has to deal with the incompleteness of other important reasoning tasks. If they are sensible, ontologies should refer to existing standard ontologies and therefore should be expressed using some standard language. A state-of the-art report on choosing an ontology language has been produced by the IST Esperonto project.[56]

7.3.1.2 Naming Conventions

Every ontology development should strictly follow one naming convention. This not only makes the ontology easier to read and to understand, but also helps to avoid some common modelling mistakes.[57] In the following, some commonly agreed conventions are listed:

- Names are expressed using the English language.
- Ontologies are identified by short, descriptive names in the singular, using lower-case letters and 'owl' as filename extension (for example 'agent.owl').
- Concepts are named with capitalized nouns in the singular, although a concept represents a collection of objects (for example 'Wine')
- Names constituting entity names are connected with an underscore as delimiter and not using the Intercap style that has been adopted in RDF and by the XML Meta Content Framework;[58] thus 'Business_meeting' rather than 'BusinessMeeting'.
- Instances and properties are named with singular nouns (rather than verbs), or sequences thereof, in lower-case letters only (for example 'chianti' and 'colleague').
- Property names should not be prefixed by 'has' or 'is_', or be post-fixed by 'of'; nor should they use the verbal forms; thus 'parent' rather than 'parenting' or 'parents'.

7.3.2 Structuring

Ontologies are employed in various fields of application. As a result, different scopes of ontologies have emerged, as exemplified by the layered architecture of ontologies developed within the IST Project WonderWeb.[59] The ontologies at lower layers provide representation requirements for the higher layers, whereas ontologies at the upper layers provided design guidelines to the lower layers.

We group ontologies into three broad categories (see Figure 7-7). Therefore, we have so-called *foundational ontologies*, which contain high-level domain-independent concepts based on formal principles derived from linguistics, philosophy and mathematics such as objects, events and processes (i.e. a broad coverage) at the highest layer. On the next layer we have *core ontologies*, which constitute the toolboxes of eminently reusable information-modelling primitives. Core ontologies provide domain-specific infrastructure (i.e. medium coverage), they specialize foundational ontologies and they help to integrate application-specific knowledge, mediating between foundational and application-specific ontologies. In that respect, a core ontology provides the infrastructure for a library of *application-specific ontologies* that reside at the bottom layer and relate concepts and properties in the domain of interest (i.e. they have small coverage).

Foundational ontologies, together with the more abstract parts of core ontologies, are sometimes referred to as *upper ontologies*. By providing repositories of standardized knowledge-representation primitives, upper ontologies foster semantic interoperability in distributed information systems. Additionally, alignment to upper ontologies can provide a solid underpinning for application ontologies and may help to exclude terminological and conceptual ambiguities that result from unintended interpretations.

We further advocate the structuring of properties into hierarchies according to their source and domain restrictions. This method is effective in grouping semantically related properties and allows the automatic verification of restrictions on the addition of new properties.

7.3.3 Development Process

There is no such thing as the unique best methodology or process for developing ontologies. This means that there will always be viable alternatives. However, in this section we will discuss a development process that is pragmatic and that maximizes the ratio between the applicability of the ontology in application scenarios and the effort that is needed to develop and maintain such an ontology.

An effective ontology development process will be iterative by necessity, further refining and evolving the ontology in each cycle. During those cycles it is of utmost importance that the ontology is evaluated using the application scenarios to ensure that its concepts reflect the reality of those scenarios, while at the same time generalizing over the different scenarios.

A development process that follows the approaches for general knowledge engineering was proposed by Akkermans *et al.*[60] and Noy *et al.*[61] This development process consists of four main phases, as depicted in Figure 7-8:

- *Kick-off phase.* This first step determines the domain and scope of the ontology, as well as the requirements for the design of the ontology. It considers the reuse of existing ontologies in that domain, if applicable. By using the application scenarios, it starts an enumerated list of important concepts, their properties and interrelations.
- *Refinement phase.* This phase formalizes the results of the kick-off phase, as well as the findings from the ontology evaluation and maintenance, in a formal description language like OWL DL. In this language, the classes and the class hierarchy are designed. This design process can be started from either the most generic concepts (top-down) or the most specific concepts (bottom-up). Domain experts play a leading role in this exercise.
- *Evaluation phase.* The evaluation phase serves as a proof for the usefulness of developed ontologies. Obviously, the ontology should be checked against the requirements defined in the kick-off phase, but, more importantly, the usage of the ontology is tested in the context of the application scenarios. This can be a whiteboard exercise; the best proof of concept is to make the ontology available to the application developers, while observing the way they use it. Collected feedback and experiences from those developers is valuable input for the further refinement of the ontology.
- *Maintenance phase.* The last phase is primarily an organizational phase, but a very important one, because it is highly likely that an ontology will develop over time and during

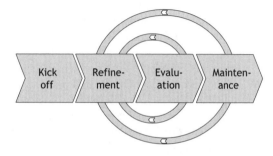

Figure 7-8.   Ontology development process

actual usage. It is therefore important that appropriate rules and tools are developed to update the ontology (delete, insert and change concepts or relations) and that the responsible parties and persons are appointed to perform the maintenance. Clear processes have to be developed to request ontology updates, for example via a web site, and to issue new versions of the ontology.

7.3.4 Standard Ontologies

Integrating huge and abstract ontologies in domain-specific scenarios is a great challenge and often an inefficient and time-consuming task, whereas small and lightweight versions of upper ontologies can provide access to a common vocabulary. With such a common vocabulary, interoperability between different knowledge representations may become feasible.

7.3.4.1 Core Ontologies

To classify the numerous existing core ontologies, it is essential to determine uniform criteria such as their type, their quantitative size and their language species. The following ontologies were identified as OWL DL upper ontologies: DOLCE (Lite) – the Descriptive Ontology for Linguistic and Cognitive Engineering,[62] the OpenCyc Spatial Ontology,[63] the REI Policy Ontology,[64] FOAF – Friend Of A Friend,[65] SOUPA – a Standard Ontology for Ubiquitous and Pervasive Applications,[66] OWL-Time – derived from DAML-Time[67] and SUMO – the Suggested Upper Merged Ontology proposed by the IEEE.[68] In addition, there exist some core ontologies that are either not directly relevant for the mobile business such as COBRA-ONT,[69] or not publicly available like CONON.[70] Furthermore, the service core ontology OWL-S[70] is discussed in detail in the next chapter.

In the following, we summarize the evaluation of the available and relevant core ontologies listed above. Other examinations and comparisons of upper ontologies exist for the US government and military domains[72] and for spatial domains.[73]

DOLCE, a first-order ontology, includes only upper-level ontologies and is therefore widely adaptable, at least its lightweight version DOLCE Lite, which is one of the top 10 most complex OWL DL ontologies. Both the OpenCyc Spatial ontology and the REI Policy Ontology can be classified as more domain-specific core ontologies than DOLCE, concentrating respectively on spatial information and security access and control issues. Despite its size (about 5000 concepts and roles), parts of the structure of the OpenCyc Ontology are often used as a basis in other projects. On the other hand, the REI Policy Ontology consists of only a basic taxonomic structure. Therefore, it can easily be integrated as a whole in different ontologies. FOAF is a fundamental description about agents. Its OWL DL version (FOAF Lite) is used in various projects, even though only primitive concepts are provided. Like FOAF, the OWL-Time ontology is well structured and its lightweight version is often referenced in a more domain-specific context. SOUPA includes a number of different other ontologies, such as FOAF, OWL-Time and the OpenCyc Spatial ontology, adopting parts of their vocabularies without importing them directly. The SUMO ontology contains the broadest and most abstract concepts as well as a comprehensive taxonomy of primitive concepts. Because SUMO has a full first-order basis, its OWL translation is in the undecidable fragment OWL Full. Furthermore, several inconsistencies in SUMO have recently been found using the automatic first-order theorem prover Vampire.[74]

7.3.4.2 MobiLife Context Ontologies

The IST MobiLife project[75] aims to bring advances in mobile applications and services within the reach of users and groups in their everyday life. One of the projects main research issues is the design of a general framework that supports the provisioning of services that are relevant to a user in a given context and accordingly adapt its functionality. Thereby, context is regarded as almost any piece of information available at the time of interaction. It is handled by a management framework that provides efficient means of presenting, maintaining, sharing, protecting, reasoning and querying context information.[76] In the MobiLife project, it has been decided to rely on ontology technology to represent, and to reason on, high-level qualitative context information. However, the context ontologies described below are not used as the main representation format for all aspects of context, as ontologies are generally weak in handling large amounts of data efficiently.

The MobiLife core ontologies provide concept descriptions of agents and spatial and temporal entities, as well as descriptions for classifying multiple devices and personal schedules. In addition, the situation component ontology defines application-specific concepts, to describe typical scenarios of mobile users. Partitioned into nine modules and written in the decidable fragment of OWL (i.e. OWL DL), the ontology defines more than 800 concepts, properties and individuals. The interrelations between the component ontologies are depicted in Figure 7-9.

The agent.owl module covers concepts for describing persons, organizations and groups. It is informed by the vCard standard[77] and the FOAF (Friend-of-a-friend) vocabulary[78] and provides vocabulary to describe a person's contact information and specifies relationships between.[79] The time.owl ontology builds on the subset of the standard time.owl ontology[67] used in OWL-S (time-entry.owl)[80] and integrates all of the qualitative relations among time intervals, as defined in Allen's algebra.[81] Similarly, the component ontology space.owl, which provides the general location vocabulary, integrates the Region Connection Calculus (RCC).[82] The other modules provide vocabulary concerning personal profiles (profile.owl), qualitative values (values.owl) and time-zone related vocabulary (timezone.owl). Finally, the domain-

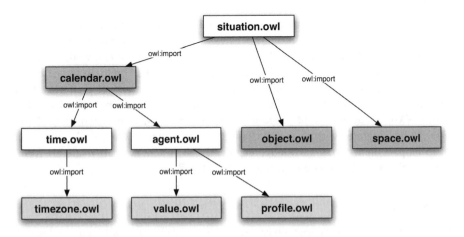

Figure 7-9. The interrelations between the component MobiLife ontologies

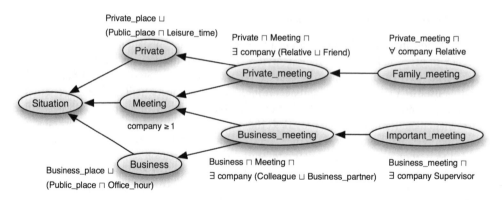

Figure 7-10.   Situation ontology

specific situation ontology (situation.owl) categorizes derived situational context specific to the MobiLife scenarios by providing axiomatically defined concepts. A fragment of the situation ontology is shown in Figure 7-10.

7.3.5 Development Environments

Because constructing ontologies is a time-consuming task, software tools have been build to support developers on most ontology engineering tasks. Current ontology development environments not only support the knowledge engineer in the creation of ontologies by providing simple editing environments, but also offer graphical ontology navigation, validation and debugging support, version management and solutions for linking to ontologies from external sources. In the following sections, we introduce the three most innovative development environments for OWL ontologies, namely Protégé,[61] SWOOP[49] and OntoTrack.[83]

7.3.5.1 Protégé

The Protégé system is an environment for knowledge-based systems development that has been evolving for over a decade.[61] Protégé began as a small application designed for a medical domain, but has developed into a much more general-purpose set of knowledge-engineering tools, which recently also included enhanced support for OWL,[84] build on top of HP's Jena libraries.[41] The current version, Protégé 3.1, can be run on a variety of platforms, supports an enhanced graphical user interface, interacts with standard storage formats such as relational databases, XML and RDF, and has been used by numerous individuals and research groups. Protégé can be integrated with several reasoning engines (such as RACER;[20] see also Section 7.1.4) to support the working knowledge engineer. The system is freely available from, and actively developed and supported by, Stanford SMI, USA.

7.3.5.2 Swoop

SWOOP (Semantic Web Ontology Overview and Perusal)[49] is a Java-based graphical ontology editor that employs a web-browser metaphor for its design and usage. It provides a hyperlink-based navigation across ontological entities, history buttons for traversal and book-marks that can be saved for later reference. As well as the RDF/XML syntax of OWL, it also supports the official abstract syntax of OWL DL.[29] Like other environments, SWOOP also comes with multiple ontology support. Validation support and species validation is realized by a tight coupling with the Pellet reasoner[21] (see also Section 7.1.4). Among SWOOP's advanced features are an extensive version management support in which change sets can be attached to collaborative annotation messages created using an Annotea plugin,[49] the possibility of partitioning ontologies automatically[85] by transforming them into separate $\mathcal{E}$-connected ontologies[86] and to run sound and complete conjunctive A-box queries written in RDQL.[36] More recently, a debug mode has been added that supports glass-box and black-box debugging by exposing the internal workflow of its DL-reasoner Pellet in a meaningful and readable manner to help users in understanding the cause of the detected inconsistencies.[87,88]

7.3.5.3 OntoTrack

The ontology authoring tool OntoTrack[83] overcomes the drawbacks concerning search and navigation speed of other editors that use two functionally disjunctive interfaces for either editing or browsing, by integrating both 'in one view'. The system combines a sophisticated graphical layout with mouse-enabled editing features optimized for efficient navigation and manipulation of large ontologies, and features an explanation component.[89] Its engine provides sophisticated layout algorithms enabling animated expansion and de-expansion of class descendants, zooming, thumbnail views and panning. Its most innovative feature is the instant feedback functionality. By synchronizing every single editing step with an external reasoner, relevant modelling consequences are instantaneously detected and visualized. Currently, OntoTrack is able to handle most of OWL Lite excluding individuals, datatype properties and annotation properties.

7.4 Discussion

The ontology technology introduced in this chapter is the result of more than 30 years of research on knowledge representation, formal logic and automated reasoning. Currently, the most appealing application of the techniques developed is the circumscription of resources in the emerging Semantic Web. For the future, however, this success may even be outperformed by its application to context-aware mobile services. In particular, the integration of decidable reasoning facilities in the mobile-service provisioning process may be of outstanding importance. In this way, the consistency of the provided service specifications can be verified and, by making use of formalized world-knowledge, implicit context information can be derived. Thus, a mobile system will be able to detect and rule out faulty service descriptions and provide reliable situation interpretation. Both aspects are important issues that need to be addressed to realize proactive service provisioning scenarios.

However, integrating ontology-based representation and reasoning support is still a complex matter, which requires far-reaching experience in various fields. First of all, ontology development still lacks efficient and reliable tool support, even if noticeable progress in this has been made recently. However, even popular ontology development tools like Protégé are still not able to provide full language support today.[61]

Moreover, while maintaining the decidability of core reasoning tasks on the one hand, OWL offers only a restricted expressiveness on the other hand, which often results in merely abstract scenario descriptions. Creating context-aware future mobile services and applications means accepting the challenge of defining preferably concrete situations. The question whether or not this can be accomplished by exclusively using OWL, while living with its somehow limited expressiveness, has not yet been answered.

8

Semantic Services

Massimo Paolucci

8.1 Challenges and Opportunities

In the last few years, the challenge of mobile computing has been to provide Internet access anywhere and at any time. This effort resulted in a great increase of connectivity for mobile users, which ultimately enabled applications such as iMode,[1] which provides mobile users with virtually unlimited access to the Web from their mobile phones. Using iMode, users can access Web pages, read the information that they provide and make decisions about their own lives.

Now that connectivity any time, anywhere is guaranteed, the new frontier of mobile computing is to enable mobile platforms to process this information to support the users in their increasingly complex life. (In this chapter the term *mobile platform* is used as a generic label indicating mobile phones, PDAs and any type of handheld device capable of performing a computation for the user.) Mobile phones have to do more than just retrieve Web pages for the user to read; rather they need to gather the information and present it in the context of the decisions that the users have to make. The problem is changing from delivering information to a mobile platform to making mobile platforms proactive participants in their users' lives by providing critical information that is needed to improve the mobile users' decision making.

This vision, although futuristic, is progressively becoming reality. Some of the services that are required to support this change are already available. For example, there are services that alert us when it is time to go to catch a flight or that the flight has been cancelled, or to tell us of the delay of a train that we plan to board. The trend toward providing services that are addressed to mobile users is only just starting, and the near future is likely to see a speed up of this process as the technology progresses and services become more ubiquitous.

But before this vision becomes a reality, many important technical challenges need to be addressed. First, the mobile platform should be able to *discover* the services that the user needs. Once these services are identified, the mobile platform needs to *invoke* them: provide

Towards 4G Technologies. Edited by Hendrik Berndt.
© 2008 John Wiley & Sons, Ltd.

the required inputs, interpret the resulting outputs and present them to the user. In general though, there will be no single service that satisfies the needs of the user; rather the available services will need to be *composed* to form more complex services, which, working together, solve the problem.

Emerging Web Services standards, such as WSDL,[2] SOAP[3] and UDDI,[4] provide a natural starting point to address the problems of services for mobile computing. Web service technology aims at defining a *Service Oriented Architecture*[5] for the Web, in which information is not exchanged through Web pages but with services. The technical challenge for mobile computing is to adapt these technologies to platforms that have limited computational power and operate in very dynamic contexts with unreliable and restricted communication. In contrast to Web Services and applications that reside on very well connected computers with virtually unlimited resources, mobile platforms reside in the pockets of users who constantly move between environments. The smart phone of a person who is going from an underground station to a surface railway station will have to adapt to a totally new computation environment: the services that were available in the underground station may no longer be reachable, and the smart phone will have to replace them with a new set of services available in the train station.

Furthermore, the dynamism of ubiquitous services in mobile computing is accompanied by a need for higher automation to free the user from the burden of managing the interoperation with such services. Whereas in a B2B context it is feasible (though expensive) to ask to a programmer to implement a client that interacts with the service, such an opportunity is not available in mobile computing. Nobody, not even the savvier user, will spend time programming a client to a service while running to catch a bus or while on an errand. The crucial linchpin of the success of ubiquitous computing will be the ability to provide clients that automatically configure themselves to operate in changing environments and that will be able to exploit the information that is provided there.

Unfortunately, Web Services standards fall short of supporting such a high level of automatism because they fail to represent the semantics underlying the transactions between the service and its clients. The main claim of this chapter is that exploiting semantic information is crucial for the future of mobile computing. Moreover, we claim that Web service technology needs to be grounded on ontologies (see Chapter 7) and on Semantic Web[6] technology, which provides the language to describe information and the relationships between different types of information through logical axioms and a well founded logic inference mechanism that together support the derivation of consequences that are consistent with that information. Ultimately, the logic-based reasoning systems that underlie the Semantic Web allow clients to decide how to use a service and how to interpret the messages that they receive from that service.

In this chapter we discuss the current state of Semantic Web Services technology, which applies Web service technology to the Semantic Web, and we discuss in some details how such a technology allows access to ubiquitous services. The outcome of this chapter will be a vision for ubiquitous services for mobile computing, which will support the level of dynamic interaction that is required by ubiquitous services. While this vision is already grounded on existing technologies, much of it has still to be realized; we will therefore present open challenges and describe existing pitfalls on the way to the realization of our vision. The rest of the chapter is organized as follows: in Section 8.2 we provide a hands-on example of the problems of interacting with already existing services. In Section 8.3 we discuss Web

Service technology and in Section 8.4 the need for semantics; in Section 8.5, we describe how Web Services and semantics are combined in OWL-S[7] to address the discovery and invocation problems highlighted above. In Section 8.6 we briefly review other Semantic Web Services technologies and, finally, in Section 8.7 we discuss open challenges. Section 8.8 gives our conclusions.

8.2 Hands-on Experience

Our vision of the ubiquitous services that will be accessible through our mobile platforms is not a futuristic idea suited only to scientific conferences; rather, it is fast becoming a reality. It is already possible to access many services through SMS or iMode, such as paying for train tickets and performs airline check-in. To understand the state of the art, in this section we analyse how a traveller can use her cell phone to receive train schedule information in Italy and in Germany. Our description is based on existing services that are provided by Ferrovie dello Stato (hereafter FS), the Italian railway company, and Deutsche Bahn (hereafter DB), the German railway company.

The first problem that the user faces is the discovery of such services. They can be found through brochures that are published by the railway companies or on the web sites of those companies, but crucially there is no 'one stop' solution where the user can query for the type of service that she needs. In other words a directory of available services for mobile platforms is still missing. (There are web sites like fastfind.com that allow one to find some of the SMS services of a company, but the company has to be known. The question 'find train schedule' fails to find any service.)

Once the service is discovered, the second problem of the traveller is the invocation of the service. Specifically, the traveller needs to compile the message to send to the service. Not surprisingly the format of the messages expected by the two railway scheduling services is different. The Italian service expects a message with the following format: 'Da Roma a Milano 27/06/2008 10:30' (http://www.trenitalia.it/it/area_clienti/sms/index.html) to access the schedule of trains leaving Rome directed to Milan on 27 June 2008 leaving after 10:30. The German service, however, expects a message in the following format: 'BAHNMIX Berlin Hamburg 27.06.08 10:30' (http://www.bahnmix.de/mix_mobil.php) to retrieve information for trains from Berlin to Hamburg for the same date and time. Critically, the two messages should not be mixed up since the German service will fail to reply to a message with the 'Da...a...' format used by the Italian service, and similarly the Italian service will fail to reply to a message starting with 'BAHNMIX...' as required by the German service. The third problem is to provide the correct arguments in the right position. The message 'BAHNMIX Hamburg Berlin 27.06.08 10:30' will provide the schedule of trains going in the opposite direction to that which the user requires. And, finally, the user will have to interpret the results of the request and extract the schedule of trains.

To make things worse, many services may require more than one message. For example, a seat reservation service will require at least three messages: the first one to find the train; the second one to check the availability of seats; and the third one to pay for the reservation. Each one of these messages will have to be carefully crafted and refer to the previous messages, forcing the traveller to remember information that could be recorded by the device.

There is clearly a problem with such services: the user should only care about the journey, where to go and the time of departure and arrival, while all other details should be taken care of by the mobile phone itself. The challenge, therefore, is how to enable such a high level of automation. Web service technology provides an initial answer to this problem.

8.3 Web Services

To support the user in her interaction with the service, the mobile phone needs to create a client that manages such interaction. The implementation of a client of a service requires a precise specification of the interaction protocol of the service. Information such as the message format and the port to which to send the message, and from which to receive an answer should be unambiguously specified. At its very core, Web Services technology provides a set of languages and specifications that support the automatic compilation of a client in a platform independent way.

8.3.1 Web Service Description Language

At the basis of the Web Services technology lies the Web Service Description Language (WSDL) (http://www.w3.org/2002/ws/desc/) for the specification of operations performed by a service and the messages required to invoke those operations. The structure of WSDL documents is displayed in Figure 8-1. WSDL adopts the view that a service is an *interface* which exposes the *operations* performed by the service. Each operation specifies a transformation between some inputs into some outputs. (In addition to input/output operations, WSDL supports the additional operations such as *input only* operations that just receive information without sending any feedback to the client, and operations that push information to the client which are defined as *output/input, output only*, and others.) The inputs and outputs parameters of the operation are defined through the message specification, where each part of the message corresponds to one parameter, and the structure of these parameters is defined as a *type* specification defined using the XML Schema Definition language. To support the invocation of the operation, each operation is associated with a binding that specifies the type of protocol, and the binding is associated with a port that specifies to where a message should be sent or from where it will be received.

Since WSDL is used to expose the functionalities of services to their potential clients, it could be used to expose the scheduling services of the Italian and German railway systems. The two services may be described as having an operation that takes as input the departure and arrival cities, together with the departure date and time, and that reports the schedule as output. The message and binding section will specify the different message formats as well as the telephone number to which to send the message.

WSDL has a number of advantages: first of all, it abstracts the service implementation details, exposing only the service functionalities. Second, it supports the specification of interfaces based on synchronous and asynchronous message passing, allowing the specification of a wide range of service types. Moreover, constraints such as platform or language or operating system or vendor dependence, which plagued other distributed computation schemata, such as Corba (http://www.corba.org/) and DCom (http://www.microsoft.com/

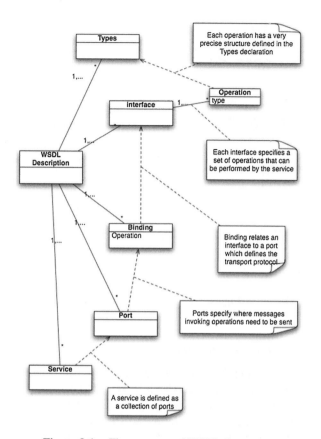

Figure 8-1. The structure of WSDL documents

com/default.mspx), do not affect WSDL. These features make WSDL very appealing for mobile computing, and indeed organizations such as the Open Mobile Alliance (OMA) provide guidance on how to use WSDL in a mobile computing environment[8] and 3GPP provides a specification of how to use WSDL in combination with SMS and other protocols.[9]

WSDL is of help in solving the problem of data exchange between the service and the mobile platform. The problem left unsolved is what to do with the data once it enters the smart phone, and equivalently how to select the data to send to the service. Strictly speaking, the use of data is outside the scope of Web Services standards, and WSDL in particular, since they concentrate on the specification of the interface. Yet, how to use the data is a crucial problem for all applications using Web Services. Generally speaking there are two ways to address this problem: the first one is to implement custom code to address the mismatch, the second one is to exploit other computational mechanisms to extract the 'meaning' of the data used by WSDL and use it to address the mismatch. In Sections 8.4 and following we will show how this can be achieved using ontologies, as discussed in Chapter 7, and OWL-S.

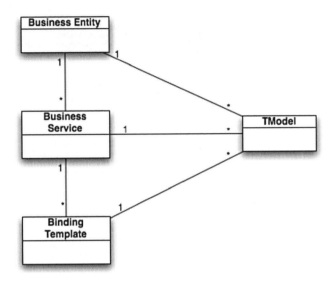

Figure 8-2. UDDI data structure

8.3.2 Universal Description Discovery Integration

While WSDL is of help during the invocation of the service, the mobile platform is aware of where the WSDL specification of the service is located. The discovery process is addressed by the *Universal Description, Discovery and Integration* (UDDI).[4] UDDI supports discovery by defining the set of standards that is required to specify a repository for Web Services. Specifically, UDDI defines the type of operations that are required to register with such a repository and to query the repository. Furthermore, the UDDI data structure, shown in Figure 8-2, provides a way to represent services so that they can be discovered. In UDDI, service providers are represented by *Business Entities*, each provider may register many services, named *Business Services* and in turn each service is associated with a binding specification. All the UDDI entities can be qualified using *TModels*, which are abstract specifications of properties that the service can assume. TModels can be used for many purposes, such as to specify the location of a WSDL description of a service[10] or to classify the service within standard service taxonomies such as UNSPSC (http://www.unspsc.org/).

The TModel representation, while very powerful, is too general. For example, two different TModels may be used to represent the same feature, while similar content may be represented by different TModels. Ideally, industry groups will provide standard industry specific TModels, removing such ambiguity, but this process has not yet started; UDDI is not therefore much of a guide in the representation of the salient features of a service.

8.3.3 Service-Oriented Architecture

UDDI plays an important role within the Service Oriented Architecture (SOA) by providing the central registry component that performs the discovery of Web Services. Following SOA (see Figure 8-3), service providers *advertise* their services with the registry, typically a UDDI

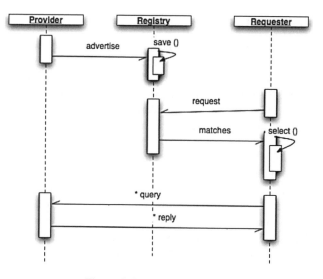

Figure 8-3. SOA Interaction

server, and then wait to be discovered by the service providers. A service requester *requests* from the registry the services that fit its requirements. The registry reports to the requester the advertisements that *match* the request and finally the requester *selects* the most appropriate provider and initiates an *interaction* with it.

SOA is an architecture well suited to Web Services distributed on the Internet, because it mimics the Web architecture by placing a search engine at the centre of all interactions. However, the centrality of the search engine is problematic in mobile computing, where some of the services may be globally available through the Internet, while others may be local, provided only in a restricted environment. If taken literally, SOA suggests that environmental changes should parallel moves from one registry to another. Yet many services may have a very limited range, only a few meters for Bluetooth-based services and centimetres for RFID, these services may not have access to any UDDI server and, anyhow, they work better using a direct P2P connection to their clients.

Web Services standards such as WSDL and UDDI provide the basic functionalities to support the interaction between mobile platforms, such as smart phones, and the services that are available. They provide standard XML-based specifications that reduce the interoperability barrier between the services and their clients, facilitating the interaction from mobile applications. However, the assumptions that they make are too strong for mobile computing, from both the representation point of view and the architectural point of view. On the representation side, they provide only syntactic interoperability and no representation of the semantic of the data exchange; and, ultimately, it is the lack of semantic interoperability that prevents automatic interoperation between the client and the service. On the architectural side, the dependence on a central registry prevents the serendipitous discovery of local services with a very limited geographical range. In the following sections we will attempt to present ways of overcoming these limitations.

8.4 From XML to Ontologies

The problem of Web Services is their dependence on XML as the communication language between the client and the service. XML is based on a well-defined syntax and therefore the service and its clients will parse the same XML document in exactly the same way, resulting in exactly the same DOM (Document Object Model). For example, consider the XML document in Figure 8-4(a), which is an XML rendering of the Italian train-scheduling service. Any XML parser would be able to recognize the structure of this document resulting in a DOM model equivalent the one shown in Figure 8-4(b).

The problem of XML is that, once the structure is recognized, there is no way to relate the document to the rest of the information of the client. Essentially, the client is left in the awkward situation of recognizing the different parts of the document, but failing to recognize its meaning. Our understanding of the words 'time', 'da', 'a', 'Milan', 'Rome' and 'date' helps us to understand that the document is about a trip from Milan to Rome at a given date and time, but in XML there is no mechanism to associate meaning with tags and therefore the meaning is hardcoded by the programmer into the application that receives the document.

The requirement that a programmer is required to deal with the meaning of data is very problematic for ubiquitous and mobile computing. Mobile users move quickly from one environment to another, change requirements and still have a crucial need of precise and up-to-date information. They have no time to stop on the side of the road waiting for the implementation of a client to a service that they may need. Rather, mobile phones need to adapt to the changing environment, discover new services and interact with them as autonomously and automatically as possible.

To remove the programmer from the loop, mobile devices have to go beyond syntactic interoperability to access the meaning of the information contained in the messages that they receive. In the case of the XML document in Figure 8-4, the mobile phone will have to know that the tags <to> and <from> define a relationship between a journey and its departure and arrival cities. In essence, the mobile platform will have to be able to abstract from the specific superficial form of the information that it receives in order to recognize how this information fits into the general context of the information available to the mobile phone.

Ultimately, the problem of XML and XML schemata is that they represent the syntax of data and not its *semantics*. This problem is addressed by the Semantic Web efforts[6] and specifically by OWL (Web Ontology Language).[11] The Semantic Web aims at describing *ontologies*, which are conceptualizations of a given domain, representing the relationships that characterize the world of the user.

Specifically, OWL makes a distinction between two types of data: the *TBox*, which is a description of the abstract relations between classes of objects and events; and the *ABox*,

```
(a)  <queryMessage>              (b)  queryMessage.da = Milan
       <da>Milan </da>                queryMessage.a = Rome
       <a>Rome </a>                   queryMessage.time = 10:30
       <time> 10:30 </time>           queryMessage.date = 27/06/2005
       <date> 27/06/2005 </date>
     </queryMessage >
```

Figure 8-4.   (a) An XML document and (b) its DOM structure

which provides a description of the objects known to the user in terms of the relationhips in the TBox. For example, the TBox may describe travel as a type of event that has a departure location and an arrival location, and distinguish train travel from other forms of travel, such as flying or travelling by car. Furthermore, it may add restrictions on the type of travel; for instance, flying requires departure and arrival airports, while train travel requires departure and arrival stations. Moreover, it may specify that train stations and airports are related to cities, etc. The ABox will contain descriptions of actual train stations, such as Milan Central station, airports, and specific journeys.

In addition, OWL provides a logic, based on Description Logics,[12] to derive consequences from the TBox and the ABox. For example, an object with a departure station and an arrival station will be automatically classified as a train travel. In addition, Description Logics inference engines can verify the global consistency of the statements that are provided to a service, so that a train journey between locations that are not train stations will be automatically recognized as invalid.

Superficially, ontologies seem to mimic the class/instance distinction that characterizes Object Oriented Programming (OOP), and that can be expressed using XML schemata. Indeed, XML schemata allow the construction of the same taxonomy structure produced in OWL, with travel as a superclass of both train travel and air travel. Furthermore, both Description Logics and OOP provide inheritance, where, if a journey has two class variables, *to* and *from*, the same class variables are shared by the subclasses.

The similarity, however, is very superficial since OOP does not provide any inference capability and therefore all the information about the known objects should be explicitly specified. Ideally, it should be possible to define a set of standard XML schemata that are shared across the Internet and that describe a common set of travel types. This is the approach followed by attempts such as EBXML (http://ebxml.org/) and Rosetta Net (http://www.rosettanet.org) and more generally by EDI standards. The problem, of course, is that the different parties may have very different viewpoints on the same object. For example, for accounting reasons the railway companies may have to distinguish between week-day and week-end travel so that the can pay overtime to engineers. On the other hand, a user may distinguish between business travel and personal travel, the first to be paid for with the company credit card, the latter with the personal credit card. In these cases, the same travel description may have to be classified in different ways on the service side and on the user side. With standard XML schemata, the additional classification will have to be hardcoded, while with OWL, the same classification can be performed automatically by an inference engine without any need for a programmer to be involved.

The automatic classification provided by OWL is crucial to supporting interoperability between clients and services in a mobile environment. Services and clients may have very different ontologies. The client may be concerned with the many different means of transportation and under what conditions they can be taken. The railway company, in contrast, may be concerned only with train transportation, but it may have a very detailed ontology specifying the personnel and machineries that are involved with a train journey. Yet, as long as services and clients share a common set of concepts, they will be able to interoperate correctly.

Ultimately, mobile services need to merge the Web Service technology with Semantic Web technology. Web Services standards provide the type of information that allows the clients to be generated automatically to provide the information at the correct port using the correct interaction protocol. On the other side, OWL provides a way to compile this information in such a way that the receiving party is able to use it appropriately.

8.5 Using the Semantic Web to Represent Services

OWL-S[7] provides a bridge between the Semantic Web and Web Services, by defining a way to annotate the statements made using Web Services standards such as WSDL and UDDI using OWL ontologies. (The discussion presented here refers to OWL-S 1.1.)

Upper ontologies provide a general conceptual framework to describe a concept or a phenomenon. Other ontologies can then be defined that refine and specialize the upper ontology, for example MobiOWLS that specializes OWL-S to describe mobile services. In a nutshell OWL-S aims at providing an upper ontology for services that answers the following questions:

- What capabilities does the service provide to its clients?
- How does the service compute its function?
- How does a client interact with the service?

These three questions are closely related to the three problems of Web Services that we described above. The first question is at the core of the discovery problem: requesters look for services that have the capabilities to achieve their goals. The second question is fundamental for service composition: a fine-grained description of the operations of the service is required to compose different services into a complex one. The third question is at the core of the invocation problem: ultimately, the client needs to know what the messages are that the service expects, what responses the service sends back and how these messages map to the semantic representation of the services.

To answer these questions, the OWL-S description of a service is separated into three main components. The *Service Profile* answers the first question by providing a description of the capabilities of the service, as well as the capabilities required from a service. The *Process Model* answers the second question by exposing the workflow of a service. The main role of the Process Model is to support interactions that require multiple exchanges by specifying what data the service needs, and what data it will report, and, more importantly, the consequences of executing a process. Finally, the *Service Grounding* provides details of how the processes exposed in the Process Model map into the WSDL operations and how they result in concrete message exchanges between the service and its clients.

The direct use of WSDL in OWL-S highlights one of the main design decisions underlying OWL-S: rather than replacing Web Services standards, OWL-S aims to enrich them with semantic information. As a consequence OWL-S can be applied in virtually any service as long as it is possible to map the inputs and outputs of the service to existing ontologies. Indeed OWL-S has been applied to provide a description of Amazon.com's Web Service,[13] without requiring Amazon to change the code of its Web Service. Ultimately, OWL-S can be used to describe any service that is directed to mobile computing.

It is possible to define an OWL-S service grounding that is independent of WSDL. This option has been adopted by Masuoka *et al.*,[14] where the authors define a UPnP grounding.[15] Nevertheless, the use of WSDL is by far the most usual case.

Since OWL-S proves to be the technical basis for our claim that the use of semantics is crucial for services in mobile computing, we will provide an overview of OWL-S. In the process we will put forward a new proposal that we name *MobilOWL-S*,[16] which describes a specialization of OWL-S for mobile computing. We will then provide a few examples of where OWL-S has been used in ubiquitous and mobile applications.

8.5.1 Service Profile

The OWL-S Profile (just called Profile hereafter) aims to describe the capabilities of a service. In OWL-S the service capabilities are characterized by the function computed by a service and by a set of additional requirements such as security constraints and quality of service. Crucially, the definition of capabilities is orthogonal to the way the service computes them. The overall idea is to describe what the service does, rather than how the service achieves it results.

The structure of the Profile is shown in Figure 8-5. Following the distinction proposed above, the Profile can be divided in two sections. In the first section are the *functional capabilities* of the service that describe what the service does. These functional capabilities, shown by the ovals on the left hand side of Figure 8-5, are described by expressing both the informational and the transformational aspects of the function computed by the service. More precisely, the view of point adopted by the Profile is that the service provides an information transformation in which the inputs are transformed into some outputs. For example, a service that reports a train timetable can be described by a transformation

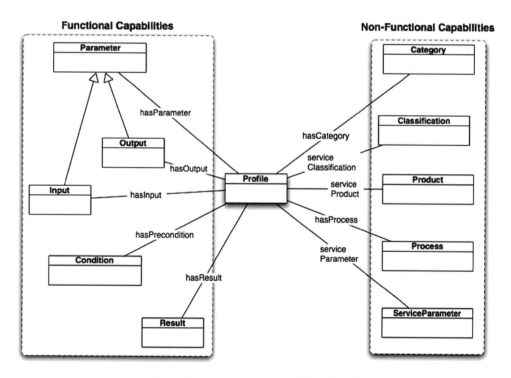

Figure 8-5. Structure of the OWL-S Profile

that, given the name of the departure and arrival cities, outputs the timetable of trains between the two cities. Other services may also produce a change in the physical world. For instance, a train ticketing service will take as input the departure and arrival cities, as well as a credit card number, and it will output an electronic ticket and the ticket price. In addition, the service will have the effect of charging the credit card account with the cost of the ticket.

These functional capabilities are not enough, since two services may compute the same function; yet the result that they provide may be very different. For instance, one of them may be very fast, the other very slow; one may require high bandwidth, the other not; one may accept Visa cards, the other MasterCard. These differences will be crucial for the user, since she may have very strong preferences toward one or the other service. The *Non-functional capabilties* of a service Profile, shown by the ovals on the right-hand side of Figure 8-5, express these properties exactly.

Non-functional capabilities are expressed through *Service Parameters,* which define a generative way of adding new features to a Profile. Specifically, for each property that characterizes a service, a new service parameter can be added. For example, service parameters can be used to specify the geographical range of a service, or its quality rating using some rating schemata. In addition, Service Parameters have been used to represent security capabilities.[17] Since, in general, the set of service parameters is potentially unbounded, the advantage of the OWL-S Service Parameters is that a new parameter can be added whenever a new feature of a service needs to be described.

In addition to Service Parameters, OWL-S provides other ways to express the features of a service. The first way is to specify the *Service Product,* i.e. the type of product that is the result of the interaction with the service. For example, a book-selling service may specify that the main type of product that it handles is books. In addition, the Profile allows the specification of the *Service Classification* and *Service Category* that specifies how the service is classified in a taxonomy of services, such as UNSPSC or some OWL-based ontology. An example of such a taxonomy is shown in Figure 8-6, which shows that there may be different types of services, such as *Commerce, Communication, Entertainment, Information* and *Travel.*

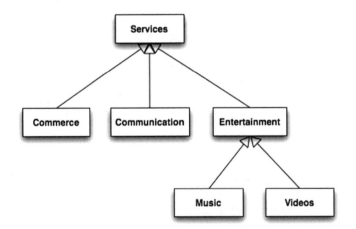

Figure 8-6.   An example of service ontology

The described service may be therefore classified within this taxonomy as belonging to one of the different classes of services.

8.5.1.1 Representing Mobility Aspects

The OWL-S Profile allows the specification of capabilities of services in terms of informational and physical transformations, but it is quite unspecific about the Service Parameters, leaving to them to domain-specific specializations of OWL-S. In this section, we propose that a new class of OWL-S Profiles should be defined that is specialized for mobile computing. This specialization, which we name *MobilOWL-S*,[16] extends the original OWL-S definition with the specification of new parameters that we found to be essential for mobile computing. MobilOWL-S extends OWL-S by providing the following four properties:

- *Comm_Channel*. This is the communication channel that is used to transmit the content between the service and the mobile platform. Using this property it is possible to specify whether the service uses Bluetooth, SMS or any other protocol, and what the restrictions are both in terms of range (Bluetooth services have a restricted range of only a few meters) and bandwidth.
- *Cost_Model*. The cost model is used to charge for the service. It may be a flat rate, a fee per use, or free.
- *Media*. This is the type of media that is used to transmit the information to the platform. For example, the media may be simple text, sound or video.
- *Device_Requirements*. The service may make some assumptions about the device that is used to display the information. Such requirements are used to specify the expected screen size and resolution, memory, processing speed and similar parameters.

MobilOWL-S provides the top-level description of service parameters for mobile services and additional subclasses can be specified for specific types of services. For example, the specification of entertainment services will also need to specify the rating of the type of content that they display and the expected audience, while location services will specify the method of location, since different methods have different degrees of precision. Specific services can be generated by instantiating the MobilOWL-S class.

8.5.1.2 Using the Profile to Discover Services

The process of service discovery requires two parties; on one side the *service provider* describes its services, while, on the other side, the *service requester* provides a description of the 'ideal' service that it would like to find. The problem of service discovery is recognizing whether the services of the provider match the description of the services needed by the requester. Service discovery is difficult because the service requester does not know what kind of services are available, nor how they are described. As a consequence, the request for services and the advertisement of services are bound to be very different even when they describe the very same type of service. Because of these differences, discovery algorithms that use string matching fail to recognize the similarity of two service descriptions. Instead, matching algorithms should abstract from the superficial description of the services so that a match can be recognized when two services are semantically equivalent.

OWL-S provides a way to perform this abstraction through the use of ontologies and their underlying logic, because the use of ontologies and logic inference allows a matching engine to verify whether the description of an advertised service is equivalent to the description of the requested service, independently of the superficial format of the service description.

A number of Web Services discovery algorithms for OWL-S have been developed in the last few years. These include the OWLS-UDDI matchmaker,[18,19] RACER,[20] SDS,[21] MAMA,[22] HotBlu,[23,24] and OWLS-MX P2P discovery[25] among many others. While they differ in the technical details, they provide a matching process that is centred on the matching of inputs and outputs. Essentially, these algorithms make sure that the outputs of the request are all matched by the outputs of the service provided, and that the inputs of the request are matched by the inputs of the service provided.

The basic type of reasoning performed during the matching process is summarized in Figure 8-7, which shows the same ontology shown in Figure 8-6. Suppose that the requester looks for a service that provides Entertainment; the search engine will start looking for services that match the request exactly. However, it will also look for services that are described by concepts that are higher up in the taxonomy, such as Service or lower in the taxonomy such as Music or Videos. Services that are defined higher up are more general than what is required and they will approximate the desired service in that they may, or may not, provide all the information that is required. The services that are below the desired one are more specific and therefore what they provide is a restrictive approximation of what desired.

The matching process provided by the algorithms described above matches only on inputs and outputs, and, to some extent, on service parameters. Future developments of discovery algorithms will need to address other aspects of discovery, such as matching also on preconditions and results of the service descriptions so that the requester can also specify what kind of conditions she would like to result from the use of the service.

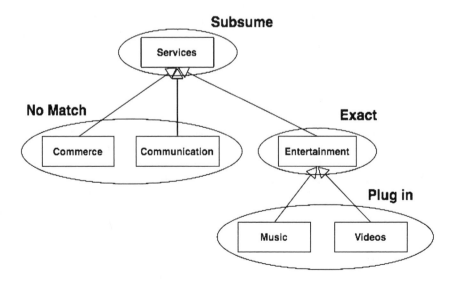

Figure 8-7. An example of ontology and degree of matching

8.5.2 Infrastructure Independence?

One important consequence of the discovery process in OWL-S, is that OWL-S does not make any architectural commitment similar to that made by SOA and UDDI. The matching process between a request and an advertisement is a functionality that can be implemented and applied in the context of very different infrastructures. Indeed, OWL-S has been applied in very different infrastructures, such as within perfect P2P networks that do not have any centralized registry and within UDDI or centralized brokers that manage the discovery as well as the interaction. Indeed, in principle, OWL-S can be used in every infrastructure that is discussed in this book.

The different discovery architectures strike very different trade-offs in terms of efficiency and applicability. For instance, the centralized broker allows very lightweight clients to interact since it centralizes many of the functionalities that would otherwise be distributed on the clients, at the cost of producing a bottleneck that may become a single point of failure, while perfect P2P architectures assume that the services and the clients need to be able to manage the discovery in addition to managing the interaction with each other, but with the gain of a very flexible and dynamic infrastructure. Ultimately, it is up to the application providers to evaluate their infrastructure and usage patterns and decide which is the most appropriate infrastructure. No matter what their decision is, OWL-S can be used.

8.6 Process Model and Grounding

Discovery leads to finding the services that satisfy the requirements of the user, but it totally ignores the interaction between the requester and the services that have been found. The interaction between the client and the service is specified by the Process Model and the Grounding. The Process Model specifies the processes that make up the service and, as a consequence, the order in which the information should be sent to the service; while the Grounding is responsible for the mapping between the processes specification and the WSDL specification of the corresponding operations.

More precisely, the Process Model is specified as a set of processes; some of these processes, named composite processes, specify control-flow statements; others, named atomic processes, specify the individual message exchanges between the service and its clients. OWL-S provides many types of control-flow statements, such as *ordered* and *unordered* *sequences*, *split* and *joints* to specify concurrent processes, *choices* to specify the non-deterministic selection of different processes and *if conditions* that specify the selection criteria among different processes. In addition, OWL-S allows a number of looping constructs to specify the repetition of processes.

An example of such a complex Process Model is shown in Figure 8-8, which shows a simplified version of the Process Model of Amazon's Web Service. The Process Model provides three choices. The first one is to browse for products and this choice will then require a decision about which type of product to browse. The second choice is to manage a user shopping cart, and here again the user needs to decide how to modify the shopping cart. The third choice is to shop, which is implemented as a sequence of the previous two operations: first the user browses Amazon's database to find the product she wants and then adds the product to the shopping cart.

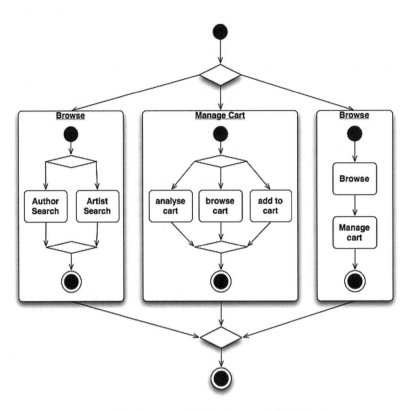

Figure 8-8. Process Model for Amazon's Web Service

In addition to specifying control constructs, the OWL-S Process Model specifies the data flow within a Process Model. The data flow describes the specification of how data is passed around between processes. For example, in the Amazon Process Model described above, the product that has been selected during the browsing operation will have to be communicated to the account management operations so that it can be added to the shopping cart. Essentially, the data-flow mechanism allows the service to publish what state information the client should keep and when to use it.

The Grounding provides a way to specify how atomic processes map to the WSDL operations and ultimately how atomic processes are invoked. The basic mapping between of OWL-S and WSDL is shown in Figure 8-9: atomic processes map into WSDL operations, while OWL-S input/output specifications are mapped into WSDL input/output messages. The latter mapping is especially important because it specifies the explicit semantics of messages. As we have discussed above, the use of XML schemata instead of ontologies prevents the use of WSDL during automatic invocation. The OWL-S mapping overcomes this problem by specifying how the XML schemata used by WSDL map to concepts defined in OWL ontologies. The mapping therefore provides a way to interpret the data that is exchanged between the client and the service.

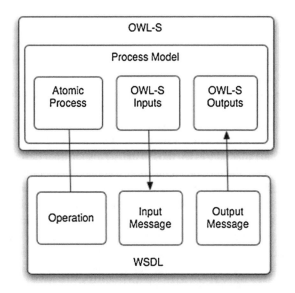

Figure 8-9. Overview of the OWL-S/WSDL mapping

8.6.1 Using the OWL-S Process Model to Interact with Services

OWL-S and WSDL provide a very powerful combination for controlling the interaction of mobile applications with services. OWL-S enriches WSDL in a number of ways. First, it provides an explicit semantics to the messages specifications so that the information that the service and its clients exchange can be specified in a way that is interpretable by both of them. Second, OWL-S specifies the order in which the WSDL operations need to be executed. For instance, the Process Model displayed in Figure 8-8 contains seven atomic processes corresponding to seven WSDL operations, but the WSDL file will not be of any help in deciding which operation to carry out next. Instead, OWL-S specifies the order in which the operations need to be executed, and the consequences of choosing a given order. For instance, it will specify that the result of browsing the book catalogue is the selection of a book and, in turn, this information will help the client to decide which is the best operation to carry out next. If the client wants to buy CDs, then browsing for books is not going to lead to the desired result.

In addition, the OWL-S Process Model also highlights what the requirements are that a client needs to satisfy toward the automatic interaction with a service. The first requirement is the ability to perform the necessary logic inference that allows the client to interpret the information that it receives from the service. The second requirement is that the client is able to navigate the process model of the service to achieve the results it wants to achieve. Essentially, the client needs an explicit description of the goals that it wants to achieve, and it needs to be able to use this description in conjunction with the input, output, preconditions and results of the process model to produce the results that it desires.

8.6.2 *Using OWL-S to Find and Interact with Services*

The starting point of the interaction between a mobile application and the services is the goal that the mobile application wants to achieve. For example, the mobile application may try to support a user who is changing flights at an airport. Guessing that the user may need to travel from Munich to Milan, the mobile application may set the goal of finding the train schedule, and buy the ticket.

Given the goal, the mobile application needs to discover a service that achieves the goal of the mobile application. The first step toward finding that service, is to compile an abstract description of the service that addresses those goals. This abstract description is compiled in the form of a Profile, where the mobile application specifies what outputs it expects, and what inputs it may provide. For instance, the mobile application may specify that it needs a service whose product is a train schedule for Italy and Germany. The second part of the discovery process is to match the abstract description of the desired service, with the descriptions of the services that are available. Finally, once all the scheduling services have been found, the mobile application can select one or more services with which to interact.

As a result of the discovery process, the mobile application will know which service to invoke. The next step is to manage the invocation of the service. This requires that the mobile application analyses the complete Process Model of the service, identifies the processes that it needs to execute and decides the order in which these processes need to be executed. In this process, the mobile application needs to infer the commitments that are produced during the interaction, so that the application does not purchase food before asking the user whether she is indeed hungry. To identify the commitments is essential because it allows the mobile application to interact with more than one service at the same time, so that it can, for example, extract the best condition possible before committing on a train or on a day of travel. Furthermore, the point of commitment provides a natural place in which to interact with the user, who may have decided not to travel – or to fly instead.

Finally, the mobile application uses the OWL-S Grounding to compile the messages to send to the service and to interpret the messages that it receives back. In the case of an interaction with the scheduling services described in Section 8.2, the application will take the responsibility for making the call and formatting the SMS messages, freeing the user from such details, and collecting data that can be used in future interactions with other services, to purchase the ticket, reserve hotels etc.

8.7 Other Proposals concerning Semantics for Web Services

In this paper we have referred widely to OWL-S as an example of a proposal to enrich Web Services descriptions with explicit semantic information. While OWL-S has been the first proposal for Semantic Web Services, and it is arguably the most comprehensive, other proposals have emerged over the years. Although it is not the goal of this chapter to review all proposed Web service description languages, for completeness we briefly discuss here three alternative proposals, namely WSMO, WSDL-S and SAWSDL, highlighting their contribution to the enterprise of adding semantics to Web Services. In addition, other proposals have been put forward, including OWL-P[26] which stresses the importance of defining policies within which the different parties can communicate, and SESMA,[27] which provides an XML syntax for semantic Web Services that maps directly to Web Services standards.

8.7.1 Web Services Modelling Ontology (WSMO)

WSMO (www.wsmo.org) and IRS III[28,29] have a vision and goals that are very similar to the vision and goals of OWL-S. Both aim at representing Web Services using ontologies from the Semantic Web, and the objective of both efforts is to support more effective discovery and interoperation between Web Services and the client. But WSMO takes a different approach; rather than stressing reasoning on the client side as the linchpin for interoperation, WSMO stresses the role of mediation in order to support automatic interoperation between Web Services.

WSMO's mediators should attack the mismatches that inevitably emerge when two Web Services, or a Web service and its clients, need to interoperate. These mismatches emerge from the use of different ontologies, or a different interaction protocol, and ultimately from the fact that the two programs have been designed with very different goals in mind.

Attacking the interoperability problem, mediators are expected to provide a solution to two important problems: reusability and scalability. Mediators facilitate reusability because the same service may be used in very different contexts by defining a set of mediators that solve the interoperation problems generated in that particular context. Scalability results from the reduction of mappings that need to be defined. Indeed, the number of direct one-to-one mappings between the different parties increases quadratically with the number of parties involved. This is very problematic because it requires a lot of implementation work. Rather, the hope is that defining mappings between mediators reduces the implementation requirements.

The idea of the mediator is very important, but also very problematic. One problem that is still underspecified in WSMO is where mediators come from. Specifically, it is not clear whether mediators emerge through logical inference, as proposed by Bouquet *et al.*[30] or whether they need to be explicitly implemented, or a combination of the two. The problem for mobile computing, of course, is that requiring implementation work to be carried out by a mobile user who is running to catch a bus is tantamount to the technology failing.

Up to now WSMO seems to be more directed at facilitating B2B transactions, rather than at supporting mobile users. This is not surprising since the initial ideas that gave rise to the WSMO proposals were based on B2B computing and Enterprise Data Interchange.[31] Nevertheless, there are new activities that have been started in the context of WSMO and aim to address the problems of mobile users.

8.7.2 WSDL-S and SAWSDL

WSDL-S[32] attempts to enrich WSDL with semantic information. The hope of WSDL-S is that enriching WSDL with semantics will facilitate tasks such as discovery of Web Services and interoperation with Web Services to ease business integration. The main aim of WSDL-S is to provide a pragmatic way to enrich WSDL descriptions with semantic information that can be used directly by programmers. Semantics is added to WSDL using 'extensibility elements', which is a mechanism to add information to WSDL that is not directly specified in the WSDL specification. Specifically, WSDL-S allows the specification of the semantics of the types used to describe the format of the data transmitted and a mapping from the data types to the semantic specification so that it is possible to extract the semantics of the data exchanged. Furthermore, WSDL-S allows the specification of preconditions and effects of operations and has one extra element that supports the specification of the type of service provided.

When compared with OWL-S, WSDL-S provides a specification that is similar to the Grounding. According to the creators of WSDL-S, this approach has a number of advantages over the OWL-S Grounding. The first advantage is that the semantics is directly available in WSDL, unlike OWL-S where the semantics is specified in the atomic process that is related to a WSDL operation. The second advantage is that it provides an approach to expressing semantics in many different ways that go beyond OWL. Ideally, a developer may also specify the semantics using UML, reducing, therefore, the need for explicit semantics.

WSDL-S evolved into SAWSDL (http://www.w3.org/2002/ws/sawsdl/), a language for the direct semantic annotation of WSDL files, and it has become a recommendation at the W3C, the standardisation body of Web technology. Since SAWSDL has reached the status of standard, other web services approaches are trying to make use of it. For example, there are already proposals on how to ground both WSMO[33] and OWL-S[34] in SAWSDL to exploit its semantic annotations. The main goal of SAWSDL was to provide a very lightweight semantic annotation of WSDL by design to increase the adoption on the industry side. To this extent, options that are adopted by all semantic web services languages such as preconditions and effects are not directly supported by SAWSDL (though it is possible to associate preconditions and effects to operations in indirect ways). One problem of SAWSDL is that the simplicity of the representation came at the cost of its precision,[35] and indeed a grounding in OWL-S requires a number of assumptions on the annotation to resolve its ambiguities. Nevertheless, by having surged to the level of standard, SAWSDL provides a first option for early adopters in the industry, and it is sufficiently lightweight for a realistic adoption in the context of mobile computing. Additional standardisation efforts will depend on the uptake of SAWSDL and on the problems that will emerge with its use.

8.8 Applications of Semantics to Web Services

In the previous sections we have proposed a service representation that addresses the problem of mobile applications which need to interact automatically with the services. In this section we provide an example of some applications that involve services with a semantic representation.

8.8.1 Using Semantic Services to Facilitate Users' Interactions

There is usually a gap between the tasks that the user wants to accomplish and the services that are available in the environment. This gap is typically solved by hard-coding some setting into machines to make them able to exploit the environment, but this is laborious work that requires expertise that only some users have and in general leads to a suboptimal use of the computational environment. *Task Computing*[14] is defined as a new computational paradigm that aims to fill the gap between users and services by exposing all services and information sources as Semantic Web Services so that they can be automatically discovered and the interaction with them can be managed automatically.

Task computing is implemented with a set of tools that form the Task Computing Environment (TCE). Most importantly in TCE are three tools: STEER, TCI and White Hole. STEER is a client that is responsible for showing the set of services that are available to unsophisti-cated users. STEER is based on Universal Plug and Play (UpnP) discovery,[15] and therefore discovers all the services that are available in a given environment. But since the environment of the user may differ from the computational environment, STEER removes all the services that are of no direct use to the user, greatly reducing the number of services that are available. The second contribution of STEER is that it compiles solutions to problems that the user typically encounters so that the user is not expected to deal with these problems.

Services available in the environment are exposed to the user through OWL-S and WSDL descriptions. WSDL adds to the services the capability to establish a connection between the user's device and the service, while OWL-S provides a way to interpret the data exchanged between the service and the device. Furthermore, OWL-S provides a way to construct the service compositions that are exposed to the user by locating the services that can consume the information that other services produce. For example, by detecting that a service provides a person's contact information, STEER can look for all services that use contact or location information, and create a composite services from the two services. Using such a composite service, the user may discover the address of a friend and see a map showng the route to that address.

Task computing makes an essential use of semantics and specifically of OWL-S. OWL-S is used to support late binding so that, instead of pre-configuring a device to work in a given environment, the device itself adapts to work within that environment. Furthermore, OWL-S is used to support the composition of services so that the capabilities that are available in a given environment are presented to the user.

8.8.2 Computing using Users' Context and Preferences

Task computing raises the problem of taking into account the context in which the user oper-ates, since context affects the solution to the problems of the user. The second problem raised by task computing is that of taking into account the users' preferences so as to find the solu-tion that more fully satisfy their needs. The representation of context and users' preferences are two problems that are actively studied in ubiquitous and mobile computing, where the representation of semantic information and the use of logic inference play an essential role. A more detailed discussion of contextual reasoning and ontologies has been provided in Chapter 7. They are also discussed here only for completeness.

Computing using context requires an explicit representation of the concepts that define the context of the user. SOUPA (Standard Ontology for Ubiquitous and Pervasive Applications)[36] is an OWL ontology that attempts to define the different aspects of context that have been defined in the literature. Specifically, SOUPA include sub-ontologies for the definition of time, personal relations through the Friend of a Friend (FOAF) ontology, space relations, policy specifications that allow the specification of constraints on resource use and so on.

SOUPA has been used in the implementation of a ubiquitous computing environment called Context Broker Architecture (CoBrA)[37] that supports mobile users in the use of smart rooms. CoBrA includes a *Context Knowledge Base*, which provides persistent storage of the context information, both in terms of general constraints such as that the same user cannot be in two different room at the same time and by recording current information on who is in the room at a given time. The Context Knowledge Base is enriched by the *Context Acquisition Module*, which gathers information from the sensors on the current state of the room and this is used by the *Context Reasoner*, which infers implicit information such as role of the people present in the room and the type of activities which they are carrying out that cannot be detected by the sensors. Finally, CoBrA uses a *Policy Management Module* to guarantee that the global policies of the room and the individual policies of the users are satisfied and that these policies do not contradict each other.

The need of semantics to represent and use contextual information is also stressed by OWL-SF,[38] which uses an OWL representation of context and the underlying inference mechanism to infer the state of a room and to enforce policies. The crucial point stressed by OWL-SF, CoBrA and other projects such as My Campus[39] is that there are a number of heterogeneous and distributed information sources that cover the sensors that are used to gather contextual information. The semantic Web Services paradigm helps to provide a description of these information sources that is immediately available to the users in a way that makes an essential use of the logic inference that underlies the semantic representation. The gathering and use of contextual information also creates the problem of how to select the service that provides the most appropriate information to the user. This is an often forgotten aspect of the discovery process, which allows selection between functionally equivalent services that provide services of very different quality. MobiOnt/MobiXPL[40] provides an initial answer to this problem by using ontologies to define a set of preference relations that reflect the users' preferences. In addition, MobiOnt provides a calculation method that exploits these preference relations to determine which is the service that best satisfies the user.

Whereas MobiOnt provides an initial approach to the discovery process, some major challenges remain, such as combining contextual information and user preferences so that the user can express that, in a given context, she has a set of preferences, but that these preferences will be different in another context.

8.8.3 Using Semantics to Control Power Consumption

The *Autonomic Networked Systems* (ANS)[41] paradigm is intended to address one of the core problems of ubiquitous computing, namely their power requirements. Computing in a ubiquitous environment typically requires extensive use of message passing. However, maintaining a connection to a network in order to receive and send messages requires energy, possibly more than a device can provide.

In the ANS vision, a service-oriented architecture should provide support for discovery of services, so that a user can find the services that satisfy her requirements. The ability to perform discovery provides the basic building block for the self-healing properties of services in ANS: upon discovering that a service is no longer available, the device automatically looks for another service that provides a similar functionality.

Self-healing, in its vanilla flavour, comes with a high connectivity cost because the client needs to monitor the presence of all services that provide a given functionality. To control such cost, ANS assumes that each service has a core functionality that specifies what the service does, and a number of ancillary functionalities that specify additional features of the core functionality. Examples of ancillary functionalities are encryption/decryption and, in general, message security, and the ability to maintain commitments on the selection of the services with which to interact.

The distinction between core functionality and ancillary functionalities has two advantages. First, it facilitates the re-use of code in the sense that the same ancillary functionality can be attached to different services providing very different core functions. Second, the use of ancillary functionalities, and specifically of commitments, allows services to control the number of messages that they exchange. Commitments are used to specify the connections that are constructed between a service and its clients and the quality of service that is provided. Relying on commitments, a client knows that it does not have to look for new services all the time; rather it can concentrate on interacting with one service.

OWL-S and, in general, semantic computation are used to support the discovery process that supports self healing, because they allow the device to express the capabilities that it needs, and the matching process required to perform soft matching and thus finding a service that provide the functionalities that approximate the needs of the device. In addition, the OWL-S Process Model is used to describe fragments of workflow that correspond to different ancillary services. With this approach, a complete description of a service may be formed by combining the descriptions of the core service and of all ancillary services.

8.9 Problems and Future Challenges

In the preceding sections we have put forward a vision for mobile computing where mobile applications need to interact automatically with the services, with no need for programmers to be involved to implement the client code. In the process, we have highlighted the contribution that is provided by Web Services standards, as well as their faults. In addition, we have tried to make a case for explicit semantics that allow the mobile application to interact with services, and we have presented OWL-S as a service description language that extends Web Services standards to provide explicit semantics. Despite the progress made with OWL-S and with other proposals such as WSMO, WSDL-S and SAWSDL, many important challenges still remain. In this concluding section, we will review the challenges and the obstacles toward a complete technology for services for mobile computing.

8.9.1 User in the Loop

Mobile computing is developing under an apparently contradictory condition: on one side users want to be in control of what their phone does and of the decisions it takes, while, on the other side, they want to be free of the technical details of invoking a service, delegating to the mobile decisions on how to transfer information and to some extent what information to transfer.

The challenge is to find a way to present the interaction between the service and the application in a way that the user is able to understand and that allows the user to stay in control of what the mobile phone is doing. This is a very profound challenge since there is a fine line between the information that can be requested from the user and the information that is already available to the machine, and which the machine may transmit directly. For example, asking the user to enter her credit card number every time there is a purchase would be very cumbersome for the user; on the other hand, sending the credit card number without any awareness on the part of the user may be even more worrisome and may ultimately result in the user failing to use the technology.

Furthermore, if the mobile phone makes a decision, it should also be able to explain why it made that decision. The mobile phone may ask the user whether she wants to use her credit card for a given payment, but it should also explain what the payment is all about, revealing all the charges that may have been obvious during the interaction with the service, but that were never disclosed to the user. More generally the user will want to know what the machine is doing.

8.9.2 Trust and Privacy

Trust and privacy are often thought of as a network problem, but there is a trust and privacy problem at the service level also. Whenever we gather information from a source, we need to gauge how trustworthy that source is. The problem will not disappear during the interaction with services.

Trust appears at two levels. First, the user needs to trust that the service, which may have been selected by the mobile application, is doing things correctly. Furthermore, the user needs to trust that the service will not misuse the information that the user provided. For example, when buying a railway ticket, the user needs to be sure that it bought a valid ticket, not a ticket for the wrong railway system. Furthermore, the user needs to be sure that the service will not use the credit card information provided to make unauthorized expenditures. Finally, the user should be guaranteed that the personal information that she provided, such as the credit-card number, will not be posted on public web sites or exchanged with third parties of which the user is not aware.

Of course, this is only partially a technological problem. On the technology side there is a need to develop a mechanism to evaluate how trustworthy services and information sources are, and to develop an overall theory for the evaluation of the different aspects of trust. However, the other side of privacy and trust lies in society, which needs to define the standards, and the relative body of laws, that will protect service users in their transactions.

8.9.3 Complete the Cycle

Despite the many proposals that have been put forward for Web service discovery, interaction and composition, there is no unifying framework yet. What many different works have demonstrated is that it is, in principle, possible to perform each one of the steps, but nobody has yet shown a complete interaction that starts from the goals of the user or application, discovers all the possible services, selects the best service, interacts with the service and uses the results to solve the original problem.

The challenge is that there is no technical solution to resolving the mismatches between the desired service and the services that have been found, nor on how to use the information about the mismatch to actually inform the interaction process about what information should be passed to the service. However, this is a very young and active field, ontologies were just a dream a few years ago, new proposals emerge at every conference and progress is very fast.

The last aspect is that it is not enough to produce the interaction; there is also a need to monitor what the client is doing and what the service is doing. The reason for this monitoring is that the client may need to explain to the user the state of the transaction; moreover, the client may also need to monitor the service to decide when and whether to interrupt the transaction.[42] For example, when buying a train ticket the client may want to know why the process is taking longer than usual, or it may need to know why an unexpected failure occurred. A number of standards have been emerging in the Web Services field concerning management and monitoring of interactions, and recently the field has been extended with semantic web techniques.

8.9.4 Using Contextual Information

An additional problem for mobile computing is that the information that is relevant will not be reliable information stored in a database; rather it will be vague and ever-changing contextual information that is collected while the user is moving. When trying to select the closest food outlet in an airport, the mobile application will need to know where the user actually is. However, location information may not be immediately available; mapping GPS information to a specific terminal may not be trivial, and GPS will not even be available in a closed environment. Other means of collecting contextual information may be as unreliable.

Another challenge of mobile computing is to be able to relate the use of a service to the information that is actually available. If location information is not available, but is required during the interaction with a service, the mobile application needs to find another way to achieve its goals.

8.9.5 Web Service Composition

In this chapter we have assumed that there is a service that addresses all the problems of the user. Although it is very difficult to provide a solution to this, it is in practice, a very unlikely situation. In general, we can imagine that no service will achieve the goals of the user; rather that the combination of a number of services may provide the service that the user needs. For example, in order to make a restaurant reservation at the airport, the mobile phone may need first to check whether the user has time to eat and then the phone may need to check whether the plane will be delayed and so on. In this case, no one service will accomplish the goal; the phone will have to use the restaurant reservation service, the departing flights service, the user's calendar and possibly other services in a very well orchestrated way.

Automatic Web Service composition is the Holy Grail of Web Service research. A number of Web Services standards are emerging that attempt to address the composition problem by producing XML-based languages, such as BPEL4WS[44] and WS Choreography,[45] which specify the complete service composition. These languages allow the programmer to specify when a service client can send a message to a service and how services will exchange information. However, the interactions that these services describe are very carefully choreographed. Indeed, these are called choreography languages and a change in the set of services used, or any change in the interaction protocol of any of these services, requires a complete change in the overall service composition which can hardly be done automatically.

There have also been attempts to compile an automatic composition,[46] and, typically, these require an automatic planner that uses means – end analysis[47] to derive which services have to be used at any given time. The problem of automatic composition is that it is not quite clear how it combines with the discovery process. Automatic planners usually start with a very well defined set of operators and attempt to combine them in such a way that the whole combination solves a problem. When they are applied to service composition, they assume that each service is somewhat like an operator, but the problem is that the mapping between the description of the results achieved by the service and the results that are required by the planner will not be a one-to-one relation. In turn, the reconciliation process requires expensive reasoning that is often computationally prohibitive.

The additional problem for mobile computing is that the goals of the user change dynamically as the user moves from one environment to another, and often the services appear and disappear dynamically. It is impossible to make a complete plan ahead of time since it is unknown which services will be available in the different environments that the user will encounter. Furthermore, the planning will have to distinguish between contingent goals that need to be satisfied in a specific situation and more general goals that the user is trying to achieve in his or her life, and make sure that the achievement of the first type of goals do not hinder achieving the latter. For example, the user may have the goal of eating while changing planes, but, if the meal takes too long and she misses her next flight, she may as well fly hungry. The satisfaction of the user goal is sacrificed for the achievement of a more important goal.

8.10 Concluding Remarks

The next frontier of mobile computing is to make ubiquitous services available to users. The problem however, is that mobile users will require an interaction mechanism that does not depend on programming clients to be able to use the services. Instead, users will require that clients are instantiated on the fly as soon as the service is discovered, and that the interaction will occur smoothly and automatically without much human intervention.

This frontier provides us with a number of technical challenges. For one, it is not enough to work with the service data structures; rather we need to find a way to express the meaning of the messages such that mobile platform can make inferences on them to decide what information to send or what the content of the information is that it has received from the user.

Web Services standards provide a way to describe the interaction between the mobile platform and the service, but they fall short in describing the meaning of the messages that the two parties exchange. Ultimately, the Web Services standards require that a programmer encodes the communication between the client and the service. OWL-S provides a way to address these problems, but there are still huge obstacles to exploiting all the power that is intrinsic in OWL-S descriptions.

The big challenge, however, is that the service technology should solve the problems of the user and that the user should derive benefits in terms of appropriate and up-to-date information from this technology, while staying in control of what information is exchanged and who receives it. Ultimately, the user possesses the key to the success of service technology, and the success does not reside in the theorems that we can prove or the prototypes that we can construct, but in the gains that the user will derive from them.

9

Dynamic Adaptation – Changing Services at Run time

Robert Hirschfeld

9.1 Introduction

We expect mobile communication systems beyond the third generation (B3G) not only to integrate several networks, but also to encourage a substantial richness of services through third-party service offerings. In this context, comprehensive support of complex and dynamic computing environments, distributed over multiple service platforms, is essential to adequately address high demands of mobile multimedia services B3G. The unanticipated nature and complexity of forthcoming services and applications makes support of dynamic service adaptation (DSA) and unanticipated software evolution (USE) a necessity.

In this chapter we give an overview of some of our research efforts towards systems B3G, including software engineering principles and mechanisms for software evolution in mobile communications systems and dynamic adaptation for service integration and personalization. We align our work with active research in the field of aspect-oriented software development (AOSD) and USE and point out how the development of highly distributed mobile telecommunication systems can benefit from the deployment of AOSD and USE.

In addition to seamless and secure access to heterogeneous networks, we consider B3G systems to encompass high service availability and best service quality to the end user. With respect to that, system requirements are highly demanding. The following are some of the key issues we have identified to be essential to B3G communication platforms:

- short development and provisioning cycles
- minimal system downtimes
- run time update and upgrade support
- third-party component and service integration

Towards 4G Technologies. Edited by Hendrik Berndt.
© 2008 John Wiley & Sons, Ltd.

- integration of heterogeneous environments
- service personalization
- context-awareness

We consider DSA to be significant in our ability to address these issues, on both the network and the terminal sides. The goal of DSA is to enable service and platform evolution, to support the advancement of individual parts at a different pace, and to facilitate personalization, context-awareness and ubiquitous computing. We consider long-lived, continuously running, highly available systems, which might be embedded or large-scale widely distributed – all which are properties of mobile communication systems – to be the chief candidates to benefit from DSA.

Adaptation mechanisms deployed these days concentrate typically on content, not so often on communication and almost never on service logic or behaviour itself. Thus, content as well as communication adaptation is understood much better than service logic adaptation. For convenience, we will use the term service adaptation to denote service logic or behaviour adaptation.

In contrast to more traditional approaches, we combine aspect-oriented programming (AOP)[1-4] with computational reflection and late binding to adapt services and service platforms when changes actually require doing so, as late as possible, preferably without disruption of service.

The remainder of this chapter is organized as follows: Section 9.2 illustrates our approach to dynamic service adaptation, addressing modularity and variation points, AOP, late binding and reflection. An overview of our research platform is given in Section 9.3. After demonstrating dynamic service adaptation applied in the context of third-party service adaptation and integration (Section 9.4), conclusions are given.

9.2 Approach

We are concerned with the what, when and how of service adaptation (Figure 9-1). The what of service adaptation distinguishes between the basic properties of software systems

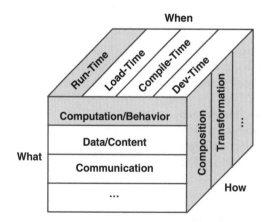

Figure 9-1.   Adaptation dimensions

computation, state and communication. The when of service adaptation addresses the time when adaptations can be made operational in a system during software development, at compile-time, load-time or run time. The how of service adaptation studies tools and techniques that allow adaptations to become effective.

The concept of adaptability is closely related to that of modularity and variation points. Modularization is a mechanism for improving the flexibility and comprehensibility of a system while allowing the shortening of its development time. Variation points allow us to explicitly designate module boundaries in a systems design where changes are expected to happen without the need of explicitly naming these changes. Variation points are introduced to support flexibility through the separation and composition of common and variable system aspects. Variations and variation points depend on the modularity mechanism provided by the programming platform on which a system is built. Most newly built systems are based on object-oriented technologies with classes and instances as modularity constructs as well as units of change. AOP provides a new, more fine-grained, modularity construct that allows us to represent cross-cutting concerns, down to the methods of individual instances.

Most changes happen after a system's initial deployment and need to be addressed very late in a systems life cycle. It is preferable, or even required, to avoid system downtimes by performing as many corrective actions on demand at run time. To address this requirement, we consider reflective architectures and late binding to be key elements of a platform for DSA.

We use the aspect modularity construct to represent units of change most adequately in our approach to DSA. Computational reflection, dynamic AOP and late binding will allow us to adapt service and service platforms when changes actually require doing so, as late as possible, preferably without system downtimes and without disruption of service.

In the following subsections we give a brief introduction to the concepts of modularity, variation points, AOP, reflection and late binding.

9.2.1 Modularity and Variation Points

Modularity is an approach to the management of complexity. By organizing a complex system into smaller less complex subsystems and then recombining these subsystems in a principled way, we are trying to improve the comprehensibility of a system as well as its flexibility and so reducing its development time.[5] We design modules that hide from each other complex design decisions or design decisions that are more likely to change.[5] Variation points, also called hotspots,[6] allow us to designate module boundaries in a systems design where we expect changes to happen without our explicitly naming them. With variation points, we gain flexibility in the context of change through the separation and composition of common and variable system aspects.

In object-oriented programming, for example, the basic unit of modularity is that of objects. Class objects capture the properties of their instances. Instead of being localized within one or a small number of modules, code that implements a particular concern is spread around (scattered) over many or even almost all places, cross-cutting various other modules, implementing other concerns as well. Because of its non-explicit structure, such cross-cutting code is difficult to reason about and, thus, is also difficult to change. The consistency of changes is both hard to prove and hard to enforce. Object-oriented programming and its class modularity construct, while proven to be appropriate for many modelling scenarios, cannot be of help in implementing the logging concern in a well modularized way.

Variations and variation points depend on the underlying modularity mechanism provided by the programming platform on which a system is built. Most modern software systems were built using object-oriented technologies where the modularity constructs, and thus the units of change, are those of classes and instances. Although this level of granularity is sufficient in some cases, a more fine-grained approach to modularity, such as method implementations, is desirable to permit the change of even smaller semantic units. Also, while traditional modules such as classes and instances may support the proper structuring of the initial system, subsequent changes to this system could cross-cut these module boundaries to affect more than one location.

9.2.2 Aspect-Oriented Programming

AOP[1-4] is a new software technology addressing the issues of separation of concerns (SOC).[7] It is based on the assumption that cross-cutting is inherent to complex systems. AOP addresses these issues by introducing new/alternate units of modularity to capture cross-cutting structures explicitly. Such structures are called aspects and can be found in a software systems design as well as in its implementation.

Aspects are units of modularity that represent implementations of cross-cutting concerns. Aspects associate code fragments (code to be executed when a join point is encountered) with join points (well-defined points in the execution of code) by the use of advice. A collection of related join points, to be addressed by an advice, is called a pointcut. Join point descriptors denote targets for the weaving process to apply computational changes to the underlying base system stated in advice objects.

The activity of integrating aspects and their advice into the base system is called weaving. Weaving in general can be performed at compile time, load time, or run time. (http://eclipse. org/aspectj/) is an example of compile-time weaving. Here, the weaver parses an AspectJ program, transforms the AspectJ abstract syntax tree (AST) into a valid Java AST,[8] and then generates Java byte code for a standard Java virtual machine. Mangler (http://javalab.cs.uni-bonn.de/research/jmangler/)[9] performs load-time transformation of Java class files. AspectS (http://www.hpi.uni-potsdam.de/swa/)[10] employs a run time weaver to transform the base system according to the aspects involved. The woven code is based on method wrappers,[11] reflection[12] and meta-programming.[13]

As of today, there are several approaches that support aspect-oriented concepts, ranging from general-purpose aspect languages such as AspectJ or AspectS to domain-specific aspect languages such as RG[14] or D.[15] Many of these languages allow us to represent cross-cutting concerns, down to the methods and instance variables level of granularity. Like objects in object-oriented programming, aspects may appear at all stages of the software development life cycle. Examples of aspects that can be commonly observed are architectural or design constraints, features and systemic properties or behaviours (such as error recovery and logging).

9.2.3 Late Binding and Reflection

During the software development and product life cycle it happens quite frequently that we find out something we wished we had known from the very beginning of the project.[16] While there is always the chance that some of the requirements were not sufficiently understood to adequately for the software system to address them, many changes happen after a systems initial deployment that are impossible to anticipate and address right from the beginning. On the contrary, such changes must be addressed very late, after deployment, during production. System downtimes can be minimized if most corrective actions can be carried out at run time.

To address this requirement, we consider reflective architectures and late binding to be key elements of a platform for DSA.

Reflective architectures are implemented by systems that incorporate structures representing (aspects of) themselves.[12] The aggregate of these structures is called the system self representation, which allows the system to both observe its own execution and influence or change its own behaviour. The former property of a reflective system is called introspection and the latter intercession. In the context of service updates and adaptation, introspection will allow us to observe computational properties of a deployed set of services, as well as the computational environment in which they are running. Intercession then can be based on our observations and result in the alteration of the service/system. While there is also research on the subject of compile-time reflection (especially in the context of generative programming,[17] we are talking about run-time reflection if it is not explicitly stated otherwise.

Late binding describes a mechanism for deferring decisions to a later point in time. With late binding, we can avoid too early commitments to design decisions, especially decisions regarding variation points that we might or will not be able to maintain. Whereas early binding requires us to provide at a very early point in time abstractions addressing possible change, late binding helps us avoid such premature abstractions. Extreme late binding allows these decisions to be made as late as possible, at run time.

9.2.4 Platform

With DSA we want services and service platforms to be adaptable, as late as possible, when changes actually require adaptation to happen, with the benefit of avoiding system downtimes and thus the disruption of service.

To carry out our research, we need to comprehend the nature of fully dynamic systems in order to advance our understanding of the possibilities, as well as the difficulties, of our ambitions. The selection and extension of our research platform is an important factor in our making progress. Our platform constituents build on top of each other, which leads to a layered architecture, as depicted in Figure 9-2.

Running bit-identically on a great variety of platforms – ranging from server machines to small devices (Figure 9-3), Squeak/Smalltalk serves us a very dynamic object-oriented multimedia scripting environment.[18] Some of its most remarkable properties are its extensive reflection support covering both introspection and intercession, its powerful metaobject protocol,[13] which gives us full access to the computational properties of our platform, and its support for very late binding to defer binding decisions until the point when they actually need to be made. The idea of metaobject protocols is that one can and should open languages up to allow users to adjust the design and implementation to make the language or environment to suit their particular needs. With that, users are encouraged to actively participate in the language design process. Language designs based on a metaobject protocol are themselves

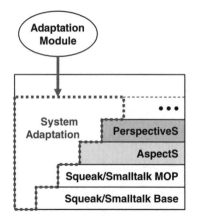

Figure 9-2. Dynamic adaptation platform

Figure 9-3. DSA platform running on a handheld device (Reproduced from reference [13] with the permission of *Journal of Communications and Networks.*)

implemented as object-oriented systems that take advantage of object orientation to make the properties of such language implementations flexible.

AspectS extends the Squeak/Smalltalk environment to allow for experimental aspect-oriented system development. The goal of AspectS is to provide a platform for the exploration of aspect-oriented software composition in the context of dynamic systems. It supports simplified guided meta-level programming, addressing the tangled-code phenomenon by providing aspect-related modules. AspectS shows great flexibility by not relying on code transformations (neither source nor byte code) but making use of metaobject composition instead. In contrast to most other approaches to AOP, which only focus on class-level aspects, AspectS allows for instance-level aspect and thus allows for modularization of behaviour that cross-cuts a set of individual instances.

PerspectiveS builds on AspectS to allow for dynamic behaviour layering in the Squeak environment. It coordinates the context awareness of a set of aspects, and so allows us to decorate a system with context-dependent behaviour, without requiring developers of the base system to be aware of potential decorations. PerspectiveS enables greater separation of concerns of a base system from its context-dependent behaviour.[19] With that, base systems can be relieved from providing behaviour that explicitly takes action in response to context changes that are not known at either development – or deployment – time. PerspectiveS facilitates role modelling by dynamic composition of multiple roles without the loss of object identity. Roles can be added or removed on demand, with each role bringing in its own set of states and behaviour.

All of these layers allow us to both implement our basic service logic and adapt this service logic to additional requirements and unforeseen circumstances if necessary. Because of the dynamic nature of our research platform, adaptation activities can be carried out on an on-demand basis, at run time, while our services are already deployed and activated.

In the next section, we use the scenario of third-party service integration to illustrate the application of our DSA platform.

9.3 Case study: Third-party Service Integration

We expect B3G mobile communication systems to be open to third-party service providers, allowing them to offer their services. Not all services to be offered will exactly match with the service platforms operated. Adjustments need to be made so that a pleasant service experience is eventually offered to customers. While some of the adjustments can be identified and applied up front, many of them will be required after the initial service deployment, perhaps without disrupting a currently active service.

In the following, we illustrate the value of DSA by discussing in more detail four such situations that can occur during third-party service integration. The situations we have selected are:

- additional safeguards
- style guide conformance
- late user-interface (UI) branding
- upgrades, updates, and fixes

This list is far from being complete. System extensions, usage indication, metering, personalization, per-standard releases and the meeting of regulatory requirements are prime candidates for being added.

All insights concerning DSA in the context of third-party service integration, safeguard introduction, style guide conformance, UI branding, upgrades, updates, fixes and so on are derived from first-hand experience, gained through prototypical implementations. Here,

Squeak, AspectS and PerspectiveS serve as part of our research platform. Squeak provides a reflective run-time environment with late binding and reflection facilities. AspectS adds quantification and obliviousness to dynamic systems. PerspectiveS offers context-aware adaptation activation mechanisms.

9.3.1 Basic Service

We decided to offer a new service called a personal digital assistant (PDA), which is intended to be used by our subscribers at their mobile terminals. The particular PDA implementation we utilize on our service platform is Fauré (http://russell-allen.com/squeak/faure/), an open-source PDA implementation designed to run on a handheld device.

Obtaining the third-party component implementing this PDA service was simple. We located the component in our component repository (the Web), and downloaded and installed it onto our service platform, so that it acted as our integration testbed. The welcome screen of an active Fauré PDA summarizes to-do items and scheduled events at a certain time for a certain day.

We can start using our new PDA right away, organizing our list of things to do, our personal schedule, our contact information or our social events to follow up. Our new PDA also offers us a little notepad where we can sketch notes or little drawings, a little piano to explore some music, 3D demos that emphasize the power of our new toy and the option just to play a game.

9.3.2 Additional Safeguards

Until now, we have faced no problem in integrating the PDA component into our service provisioning environment. Ending the PDA service by pressing the Quit button, however, will reveal an assumption made by the original developer of this third-party component that is not acceptable for our platform: Instead of ending only the PDA service, quitting the PDA will also quit the PDAs execution platform and with that our entire service platform.

To make sure that something like that will not happen in a production environment, we need to adjust the behaviour of the Quit button functionality. Instead of asking the original Fauré developers to change their components to fit our needs, and also instead of us performing those changes in the component source code ourselves, we decide to perform an adaptation in a non-invasive manner (meaning that it does not affect the original source code of the original implementation) by applying DSA. We provide an adaptation module (an aspect) that will accompany the original component and instruct our run-time environment to insert additional behaviour into the Quit button functionality, so that every time customers want to end their new PDA service they will be asked if they want to exit only the PDA or the whole run- time platform (Figure 9-4).

In this example we introduce an additional dialogue to visualize better the change applied. In a commercial system we would most likely not offer such an option, but exit the PDA only without giving the choice to exit our platform (here Squeak) in the first place.

The listings in Figures 9-5 and 9-6 illustrate how the adaptation was achieved. Figure 9-5 shows the method that is invoked every time a customer presses the Quit button: Our PDA saves its current state and, after that, invokes the quit primitive (`Smalltalk quitPrimitive`) of our platform, with the consequence of terminating the entire platform (Squeak).

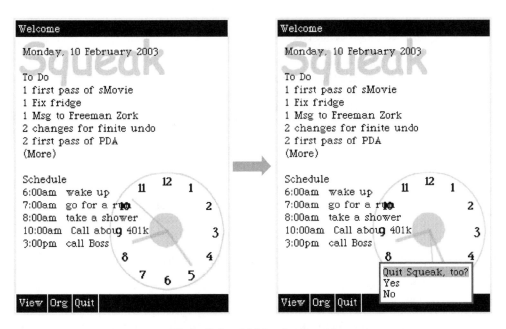

Figure 9-4. Additional safeguard

FaureWorld class>>
quit

PDA current saveDatabase: 'db.pda'.
Smalltalk quitPrimitive.

Figure 9-5. Quit primitive invocation

Figure 9-6 shows the part of our adaptation module (`FdsaQuitAspect`) that instruments the previously discussed method. Employing AspectS (http://www.hpi.uni-potsdam.de/swa/),[10] we construct an advice (`AsAroundAdvice`) that provides code to be executed instead of the original `quit` method of `FaureWorld`: After saving the state of our PDA as in the original implementation, we insert a dialogue (`self confirm:`), asking our customer if only the PDA is to be terminated or actually the entire platform (Squeak).

Our deployed PDA service is accompanied by this adaptation module, which instructs our service platform to carry out the desired adaptation step.

9.3.3 Style Guide Conformance

Many operators require third-party services provided through their infrastructure to conform to specific user-interface (UI) style guides. Prominent examples are style guides for i-mode by NTT DoCoMo and for Vodafone live by Vodafone. Non-conformance to such style guides

```
FdsaQuitAspect>>
adviceBrowserBuildMorphicSystemCatList

    ↑ AsAroundAdvice
        qualifier: (AsAdviceQualifier
                        attributes: { #receiverClassSpecific. })
        pointcut: [OrderedCollection
                with: (AsJoinPointDescriptor
                                targetClass: FaureWorld class
                                targetSelector: #quit)]
        aroundBlock: [:receiver :args :aspect :client :clientMethod |
                    | ctx morph |
                    PDA current save Database: 'db.pda'.
                    (self confirm: 'Quit Squeak, too?')
                                    ifTrue: [Smalltalk quitPrimitive]
                                    ifFalse: [self deleteFaureWorld]]
```

Figure 9-6. Safeguard dialogue for quit

can cause misunderstanding on the users' side and ultimately mean that the service portfolio offered by a service provider is not well selected. This can and will cause harm to customer acceptance, at best of an individual service offering or at worst of the entire service portfolio. Style-guide-related adaptations may not only be necessary for third-party components not developed originally with a specific style guide in mind, but also when existing style guides or policies are changed.

In our example, we have chosen a style guide that requires the text that appears on Quit buttons to be rendered using the colour red. The developers of the Fauré PDA did not anticipate changing the colour of the Quit button text to be a concern for us and, because of that, they did not provide a means to change it. Instead, the colouring of the Quit button is hidden somewhere in the UI initialization sequence of the PDA component.

Figure 9-7 shows Fauré's UI after it has been adapted by us. We applied a non-invasive adaptation module that changed the colour used to initialize text of the Quit buttons to be red.

In Figure 9-9 we can see part of our adaptation module (FdsaQuitButtonMi-grateAspect) that instruments Fauré's menu bar button construction method (Figure 9-8) as follows: We create an AspectS advice (AsBeforeAfterAdvice) that provides code to be executed after each invocation of the addButton:withAction:target: method of FaureMenuBar. Our code checks if the button constructed actually is a Quit button and, if it is, it changes its text colour to red (m color: Color red).

While such a change only becomes effective during the start-up of a PDA component, this style-guide-related adjustment also needs to be applied to all running PDA components, which means to PDA components that had already been started and had already run their UI initialization sequence. To achieve that, we greatly benefit from the reflective nature of our run-time platform: We employ a meta-program that finds all active PDAs not yet conforming to our style-guide requirements and transforms all places where it is necessary to make all existing Quit buttons render their text in red.

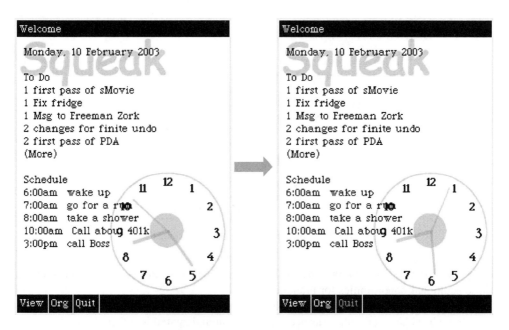

Figure 9-7.   Imposed style guide

FaureMenuBar>>
addButton: aName **withAction:** aSymbol **target:** aTarget

```
    | m |
    (m ← SimpleButtonMorph new) label: aName;
            borderWidth: 0;
            target: aTarget;
            actionSelector: aSymbol;
            actWhen: #buttonDown;
            cornerStyle: #square;
            color: Color black;
            height: 20;
            vResizing: #rigid;
            hResizing: #rigid;
            layoutInset: 3;
            changeTableLayout.
    (m findA: StringMorph)
            color: Color white.
    self addMorph: m.
```

Figure 9-8.   Button initialization

9.3.4 Late UI Branding

Many third-party components offer UI elements that could be used for additional branding, which may be used by the operator of a service platform or the service providers themselves to place brand names, trademarks or even advertisements. Unfortunately, most of the time,

```
FdsaQuitButtonMigrateAspect>>
adviceFaureMenuBarAddButtonWithActionTarget

    ↑ AsBeforeAfterAdvice
            qualifier: (AsAdviceQualifier
                            attributes: { #receiverClassSpecific. })
            pointcut: [OrderedCollection
                    with: (AsJoinPointDescriptor
                                targetClass: FaureMenuBar
                                targetSelector: #addButton:withAction:target:)]
            afterBlock: [:receiver :args :aspect :client :return |
                        | m |
                        m ← receiver submorphs first findA: StringMorph.
                        (m notNil and: [m contents = 'Quit'])
                                ifTrue: [m color: Color red]]
```

Figure 9-9.   Specialized quit button initialization

component providers do not provide explicit interfaces that would allow us to make use of those additional opportunities for branding.

DSA allows us to augment basic UI rendering to place additional branding-related information onto UI widgets and other surface areas without anticipated interfaces that allow us to do so explicitly.

The Fauré PDA comes with a 3D demo to show the high-performance 3D rendering capabilities of the Squeak environment that Fauré makes use of. The demo displays a cube with each of its six square sides rendered in different colour. Wheel controls allow this cube to be zoomed and rotated in all three dimensions.

Since the surface area of the cube is rendered using a plain texture, it is a prime candidate for additional branding. We provide an adaptation module that places an additional texture, the DoCoMo Euro-Labs logo, on its surface (Figure 9-10). The application of our adaptation can again be characterized as dynamic and non-invasive because it can be applied and revoked at run time, and it does not change the source code of the original component to adapt to our current needs.

Figure 9-11 shows the code used to initialize Fauré's 3D demo scene. Here, a 3D scene object (a cube) is created and added to the actual scene, without providing any specific texture to be rendered on the sides of the cube.

In Figure 9-12 we have another adaptation module (`FdsaDcm13dMigrateAspect`) creating an advice (`AsBeforeAfterAdvice`) that adds some code to be executed after the creation of the 3D demo scene in `createDefaultScene`. Here we provide the 3D demo object with our DoCoMo Euro-Labs logo as new texture.

This particular adaptation is yet another example for the need of instance or state migration necessary to adjust existing objects having a state that is the result of side effects that have occurred before the activation of our adaptation module.

9.3.5 Upgrades, Updates and Fixes

By looking at the logo of DoCoMo Euro-Labs in Figure 9-10, we discover a 3D rendering bug. This rendering bug is not a bug introduced by Fauré, but was one already there in our

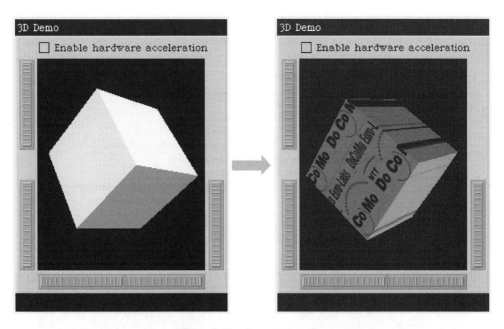

Figure 9-10.   Late UI branding

B3DSceneMorph>>
createDefaultScene

    | sceneObj camera |
    sceneObj ← B3DSceneObject named: 'Sample Cube'.
    sceneObj geometry: (B3DBox
          from: -0.7@-0.7@-0.7 to: 0.7@0.7@0.7).
    camera ← B3DCamera new.
    camera position: 0@0@-1.5.
    self extent: 100@100.
    scene ← B3DScene new.
    scene defaultCamera: camera.
    scene objects add: **sceneObj**.

Figure 9-11.   Sample 3D scene creation

run-time environment. Now that we discovered it, it would also be nice to fix it immediately, without the need to rebuild the whole system, shutting down all nodes that need to be fixed, replacing the old malfunctioning system with the newly built one and bringing everything up again. Note that bringing a system down and up again may require us to back up and restore operational state if necessary.

Instead of exercising the procedure of rebuilding and exchanging the system, we provide a dynamic adaptation module that fixes the 3D rendering problem while our system is running (Figure 9-13).

FdsaDcml3dMigrateAspect>>
adviceB3DSceneMorphCreateDefaultScene

   ↑ AsBeforeAfterAdvice
        qualifier: (AsAdviceQualifier
               attributes: { #receiverClassSpecific. })
        pointcut: [OrderedCollection
               with: (AsJoinPointDescriptor
                     targetClass: **B3DSceneMorph**
                     targetSelector: **#createDefaultScene**)]
       afterBlock: *[:receiver :args :aspect :client :return |*
                receiver scene objects first
                     texture: ((Form fromFileNamed: 'dcml.jpg')
                         asTexture wrap: true)]

Figure 9-12.   Offering a texture for the 3D cube

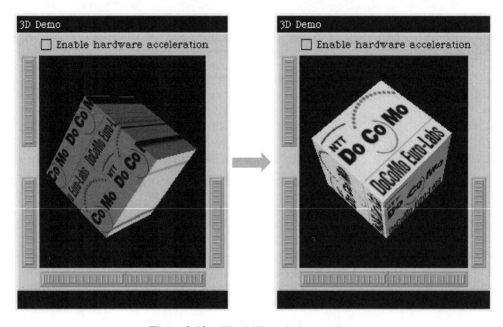

Figure 9-13.   Fixed 3D rendering problem

9.3.6 Service Integration

The Fauré PDA also provides a game called Same Game, originally written by Eiji Fukumoto for UNIX and X. The object of SameGame is to maximize the score by removing tiles from the board. Tiles are selected and removed by clicking on a tile that has at least one adjoining tile with the same properties.

But what if most of our customers would like to play another more popular game, a game like Tetris? Tetris was originally developed by Alexey Pazhitnov and Vadim Gerasimov on an Electronica 60. In Tetris, regularly-shaped blocks appear at the top of the screen and advance steadily down a fine grid. These blocks can be spun to make them fit into point-scoring rows. As levels get completed, Tetris gets faster, which makes it harder to spin and fit blocks together to complete the rows.

After searching for an implementation of Tetris, we find one that runs in our execution environment. Unfortunately, that implementation does not fit into our PDA: The UI element representing the game is too large because its height exceeds the height made available for user applications by the PDA. Also, the game control buttons that allow us to rotate and drop Tetris pieces are in a location that would cause us to waste even more screen real estate that we cannot afford.

A common approach to make the new Tetris game fit into the PDA environment would be to obtain its source code, change this source code and completely rebuild the game application. Another way to make Tetris conform to our requirements is to provide an additional piece of software that instructs our run-time environment on how to transform this game to become deployable within our provisioning environment.

Figure 9-14 shows both the original Tetris application previously discussed and the same application after its transformation and integration into the PDA. One can see how its size has been changed to meet the constraints imposed by the PDA. Also, all the game control buttons, which were previously found at the top of the game area, are now arranged in the bottom row of the PDA UI, where one would have placed them in the first place if the game had been designed to run in the PDA from the beginning.

Making Tetris fit is not enough does not in itself justify the claim that its integration is done. It needs to be accessible by the user, too. For that we have to extend the launch menu

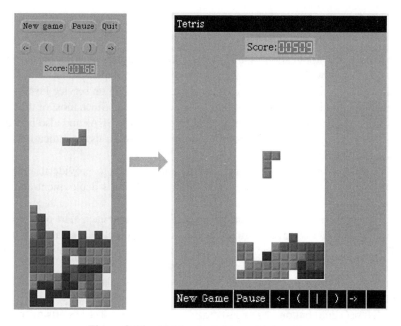

Figure 9-14.   Tetris adaptation and integration

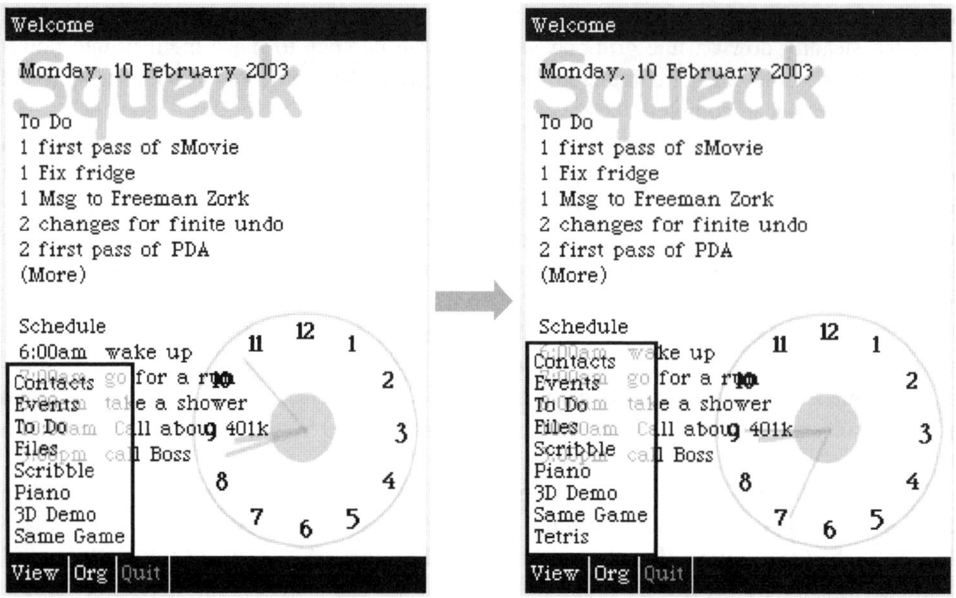

Figure 9-15. Fauré view menu with a Tetris menu entry

of our PDA by providing an entry that will launch Tetris if selected. Figure 9-15 shows the
extended menu, with our new Tetris entry last in the list.

9.3.7 Usage Indication and Metering

Merely providing new applications and services to our customers may not be sufficient from
a business point of view. Providing services most of the time implies some form of compensation, either direct or indirect. Compensation is typically based on service level agreements
(SLAs) about quantitative information on the usage of a service. Since most of the time third-
party software components are not developed to target specific SLAs and also because SLAs
can change very often, it is not of benefit to commit to specific usage indications too early
in the service life cycle.

DSA allows us to instrument our applications and services to provide usage indication
information, not just after their development, but also after their deployment and as late as
at run time.

Figure 9-16 shows a usage indication trace for Tetris, where each start of a new game is
reported to usage collection mechanisms that can act as an input feed to a rating and billing
engine. This usage indication record generation was introduced by one of our adaptation
modules that instrument the original Tetris component.

The listings in Figures 9-17 and 9-18 illustrate how this adaptation was achieved. The
first listing (Figure 9-17) shows the method that is invoked every time a customer
presses the 'New Game' button (`TetrisBoard>>newGame`): Tetris starts over with a new
game.

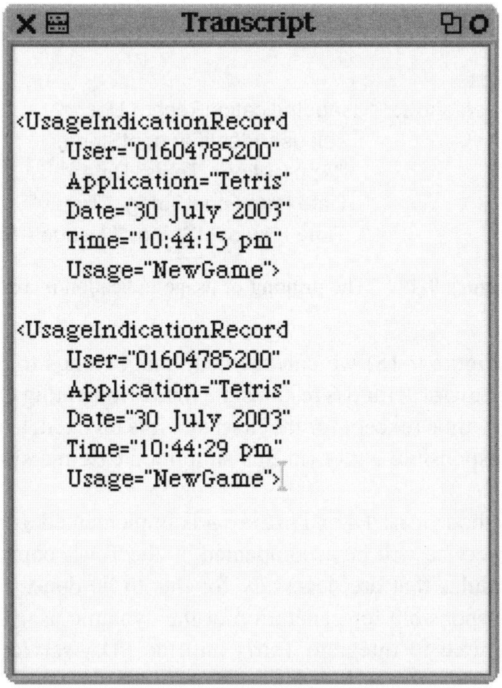

Figure 9-16. Posted Tetris usage indication records

TetrisBoard>>newGame

```
self removeAllMorphs.
gameOver ← paused ← false.
delay ← 500.
currentBlock ← nil.
selfscore: 0.
```

Figure 9-17. Starting-up a new Tetris game

FdsaTetrisUsageAspect>>adviceTetrisBoardNewGame

```
↑ AsBeforeAfterAdvice
    qualifier: (AsAdviceQualifier
                attributes: {#receiverClassSpecific. })
    pointcut: [OrderedCollection
                with: (AsJoinPointDescriptor
                        targetClass: TetrisBoard
                        targetSelector: #newGame)]
    afterBlock: [:rcvr :args :aspect :client :return |
            thisContext baseSender baseSender selector
                ~~ #initialize "the first game is for free"
                    ifTrue: [self postTetrisUsage]]
```

Figure 9-18. Posting Tetris usage indication records

FdsaTetrisUsageAspect>>postTetrisUsage

```
Transcript
    cr; show: '<UsageIndicationRecord User="',
            self userIdentifier printString,
            '"Application="Tetris" Date="',
            Date today printString, '"Time="',
            Time now printString, '" Usage="NewGame">'.
```

Figure 9-19.   The printing of usage indication records

In the next listing (Figure 9-18) we can see code that belongs to our adaptation module (FdsaTetrisUsageAspect) and is responsible for instrumenting the newGame Method in such a way that, every time (except for the first) that it is invoked, a usage indication record will be posted to the responsible entity (in this simplified case the system transcript, Smalltalk's console).

The convenience method postTetrisUsage is implemented as shown in Figure 9-19.

Our deployed PDA service will be accompanied by the Tetris component to be integrated and the adaptation modules that are necessary for this to be done. The adaptation module shown above is only responsible for generation of the dynamic usage indication record; the adaptation module required to integrate Tetris into the PDA service is not shown in this chapter.

9.4 Context-oriented Programming – An Outlook

Context information plays an increasingly important role in our information-centric environment. Context-aware applications and services are going to become the differentiator for each service provider. A wide variety of new services, ranging from those that are location-based through the situation-dependent to the deeply personalized, will provide a major competitive advantage for operators in a market saturated with flat-fee infrastructure expectations.

To date, numerous activities in developing context-aware services employ rather time-consuming and error-prone processes to eventually produce fairly constrained and mainly dedicated solutions. In this situation, such solutions both have a long time-to-market cycle and are usually hard to build upon for future context-aware services because of their specificity.

By identifying the underlying principles of context-aware services and their explicit representation and introduction into our software platform infrastructure, we believe hat it will be possible to lower the entrance barrier for both service providers to develop context-aware services and for operators to offer such services. If there is adequate support at both development and deployment times, these lowered entrance barriers suggest that we can expect an increased number of such context-enhanced services, leading to service portfolios that are extended in meaningful ways, and thus to a growing customer base, increased customer satisfaction and improved customer retention.

Our approach to achieving that goal we call context-oriented programming, or COP.[20] Together with our academic partners, we will establish a notion of context-oriented programming that goes beyond traditional programming paradigms. There is already a tradition of research in several fields of computer science that partially covers notions of adaptivity and adaptability that take context-sensitive information into account. However, our approach does not focus on application-specific solutions, but rather on the development of a programming paradigm and supporting programming language constructs and software infrastructures that improve the maintainability, robustness and reusability of software that must fit into highly dynamic environments. In this way, we will enable both service providers and operators to express context-oriented features of a wide range of services and applications in a concise and convenient way, without being distracted by unsuitable technicalities.

9.5 Summary

We believe that next-generation mobile communication systems will be more complex than ever before. Not only is this caused by the increased complexity of the environment to which these systems are connected, but it is in part also due to the assumption that such systems will be open for third-party service providers to offer their services to end customers using an operator's communications platform. Service providers come and go, their service portfolio is adjusted to the needs and the preferences of their customers, and service level agreements will be one of the key differentiation factors. Because changes are the norm rather than an exception and usually cannot be planned for, we need new concepts, mechanisms and technologies to support change better so as to make all this possible. We need to investigate what is really required to make us comfortable with change. We then have to understand either how to evolve our computational platforms to meet our needs so that we can migrate to a different and better platform or, if such a platform does not exist yet, how to build one ourselves. As well as concepts, mechanisms and technologies, we also need appropriate infrastructure support to propagate adaptation modules, to coordinate their activation and deactivation, to detect and resolve conflicts if necessary, and to address safety and security concerns related to mobile code. Change activation and deactivation, or adaptation composition in general, can go beyond a basic approach towards semantic-based service composition. With the support of COP, we will enable the expression context-oriented features of a wide range of services and applications in a concise and convenient way. We think that our research will give us a more principled approach to DSA.

Part III

Services and their Intelligent Embedding in the Environment

10

Context-aware Mobility Management

Christian Prehofer

Context awareness will be a major enhancement for future mobile systems. This holistic approach aims to adapt the communication system to the current environment of the user. In this chapter, we focus on context-aware mobility management, including handover and paging as a specific instance of this problem.

To optimize services in a heterogeneous mobile network environment, the use of context information is essential. While second-generation mobile networks provide a very homogenous network, both in terms of network topology and network services, this will change toward a more heterogeneous environment in future mobile networks. Network services (multimedia communication, high-bandwidth data services) will be available at different access points in different ways. Thus, providing optimal services to the user will be more challenging for several reasons:

- There will be more diverse radio access networks (for example wireless LAN, second- and third-generation cellular networks and their variants, and other upcoming technologies such as ad hoc networks).
- There will be many more options concerning network services (for example QoS, security, charging, roaming, etc.).
- Applications and user preferences evolve fast and will require optimal support from the network services.
- Advanced location information and group mobility information (for example users travelling in a car) will be available to support the services.

We propose a general framework for a context-aware mobility management. We will examine how to gather the specific context-aware information that is relevant to a handover. In addition, we have to cope with the complexity and profusion of information available to

Towards 4G Technologies. Edited by Hendrik Berndt.
© 2008 John Wiley & Sons, Ltd.

the mobile node. The problem is that this information is distributed over several nodes of the network (for example location server, user profile database and access points) and the mobile node, including the application layer, and must be provided to the right nodes at the right point in time. Furthermore, context information usually evolves much faster than the network layer functions. We argue that we need flexible software technology, including software agents or active networks, in order to cope with the changing nature of context information.

We show that mobile network services can also be improved by using information about the context surrounding mobile devices and networks. These context-aware network services can dynamically adapt their behaviour to the surrounding context. This includes context information related to the user, mobile devices and networks.

One of the main issues is that context information can be retrieved from a wide diversity of sources, such as user profiles, location systems, traffic monitors and sensors. The challenge in mobile networks is to collect data from these diverse sources, to process that data into context information and to distribute the processed information to several different types of context clients. These clients may be running on distinct mobile or fixed devices. At the same time, the approach has to scale to a high number of information sources, applications, mobile devices and networks.

While most of the work on context awareness is intended to support applications, we focus in the following on enhancing more basic network layer functions of the mobile network using approaches such as mobility management and paging. We will discuss which particular problems have to be solved at the networking infrastructure level. Networking differs from the application layer in several ways. First, the networking functions are usually longer lived than applications and are optimized for specific networks. Second, networking functions are highly optimized for real-time operation, as is required for an interactive communication system. In this sense, context information management has to be highly efficient. Furthermore, mobile networks have to cope with fragile and sometimes scarce wireless resources. Future mobile environments will encompass heterogeneous networks and services.

In the next section, we show examples of context-aware networking and then discuss the problems of context management in mobile networks in Section 10.2. In Section 10.3, we present different options for context management in these examples. We will show that a flexible approach that enables the installation of new software modules, 'agents', nicely fits our requirements. As a generalization of this, we introduce a general-purpose context management framework in Section 10.4. The framework provides several mechanisms for optimizing context management for networking services. In particular, we distinguish between generic, application-independent, and application-specific context handling. The last of these is highly optimized for one application. For flexibility, we support the on-demand deployment of application-specific modules.

In Section 10.5, we describe a scalable architecture that implements the proposed context management framework in mobile networks, as introduced first by Wei et al.[1] In order to deploy application-specific services in a scalable fashion, the proposed architecture includes a service deployment infrastructure able to install and configure customized modules in programmable network nodes. For performance in the network elements, a programmable platform enables the implementation of specific modules on the network layer in an efficient way. First evaluation results for this architecture are discussed.

10.1 Context-aware Mobility Management

In this section, we introduce some examples for context-aware networking functions. These serve to illustrate the main requirements of our targeted context management functions and will be used throughout this chapter.

10.1.1 Context-aware Handover

In the following, we describe the example of context-aware handover. We show that handovers can be optimized by information about the user's context. The goal is the optimal selection of a new access point (AP) during a handover, as shown in Figure 10-1.

Signal strength and the availability of radio resources are common types of information to be considered for a handover. However, they are not enough to provide efficient handovers. Even if one access point is slightly better regarding these local measurements, the decision based on them may not be the best.

For instance, in the scenario of Figure 10.1, a node moves into an area covered by both AP2 and AP3. With correct location prediction or context information, the right choice can be made. For instance, if the node is on the train, it is better to handover to AP3 or AP4, even if AP2 has a stronger signal for a short period of time.

In many cases, the handover can be optimized through knowledge of node movement and user preferences. For instance, if the node is in a car or train, its route may be constrained to certain areas. Also, the node profile may contain information that the node is located in a car. Alternatively, the movement pattern of a node may suggest that the user is using a specific type of transport.

A significant problem is that handover decisions have to be executed fast. However, the node profile and location information is often available on a central server in the core network. Retrieving this information may be too slow for handover decisions. Furthermore, the radio conditions during handover may be poor and hence limit such information exchange.

Another problem in future networks is that scanning for access points of different radio technologies can be expensive (with respect to computation power and energy consumption) or even impossible for devices that only support one mode at a time. In this case, context-aware assistance of the network to avoid unnecessary search for access points can be very valuable. With knowledge about the possible access points, the number of reconfigurations to another mode can be minimized.

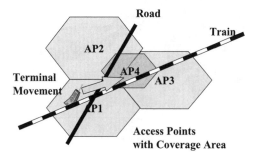

Figure 10-1.   Context-aware handover (from Prehofer *et al.*[2])

The idea of the solution is to proactively prepare the context information, such that this information can be used for handover decisions. A typical example is information about the current movement pattern (for example from knowledge of train routes or roads). For implementation, different algorithms can be deployed on the node, depending on the context information. This implementation needs a cross-layer interface that collects the context information from different layers and makes an optimized decision about deploying a decision algorithm.

10.1.2 Customized Paging Service

With paging systems, the registration update of the location of mobile devices needs to be done only when the idle device moves to a different paging area. This is illustrated in Figure 10-2. Selecting the right size and shape of the paging area is essential to guaranteeing the efficiency of the paging system. On the one hand, large paging areas increase the cost of the paging process. On the other hand, small paging areas increase the rate of registrations and the battery consumption. Current systems use a paging area of fixed size, which is not optimal in many cases. The proposed context management architecture facilitates the customization of the paging area, thus improving paging performance.

In mobile networks, the network needs to know the location of the mobile node in order to maintain the connection to it. This requires the mobile node to continuously send location updates to the network. A fast-moving node has to send updates frequently, which causes considerable signalling overhead. For this reason, paging is used instead to save the energy and decrease the signal overhead of location updates when a node is idle.[4]

With paging, the complex location registration to the networks needs to be performed only when the idle node moves to another paging area. One paging area consists of several cells with access points. In order to receive an incoming call, the paging process is used to find the exact location of the node in the paging area. The paging strategy can be 'Blanket Polling' or 'Sequential Paging'. In the first strategy, a paging request is sent to all the wireless APs in the paging area simultaneously and the AP in the cell where the node is located replies to

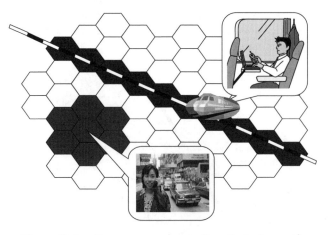

Figure 10-2.   Context-aware paging (from Prehofer *et al.*[3])

the paging request. In the second strategy, paging requests are sent to the APs sequentially in decreasing order of the likelihood that the node is located in that cell.

As mentioned above, the optimum size and shape of the paging area are essential for paging efficiency. Active network technologies facilitate the customization of the paging area. Similar to the customized handover, customized paging uses the user profile and mobility information to dynamically adjust the paging area.

Figure 10.2 gives an example of customized paging areas. Suppose the node is on the train and the direction and speed of its movement can be determined. In this case, the paging area outside the train route is not necessary. The optimized paging area should be along the railway track, as shown in Figure 10.2. Because trains may move very fast, a large paging area size is preferred to avoid frequent location registrations. Otherwise, if the node's movement is slow and unpredictable (for example a pedestrian's mobile phone), the optimal paging area should be centred around the node and should have a smaller size (because the registration cost is low). From the above examples, we can see that it will be more efficient to use an adaptive paging area according to the mobility information of the node.

There is some existing literature on the optimization of paging areas.[5] Parameters such as speed and direction are computed based on the samples received. Then the paging area is calculated from the movement parameters. In Wu *et al.*,[6] a behaviour-based strategy is proposed to estimate the mobile's location by collecting the mobile's long-term mobility logs. A similar approach is discussed by Lei *et al.*[7] An active networking solution provides the flexibility to load different algorithms. The input parameters can be changed dynamically according to different user profiles. Obviously, interaction between the network layer and the application layer is needed in this application as well.

10.2 Context Management in Mobile Networks

This section discusses the problems inherent in the collection, pre-processing, distribution and usage of context information to enhance network services. In general, context information can be static or dynamic and can come from different network locations, protocol layers and device entities.

The goal is to use the right context information at the right place and time to optimize the handover to a new access point and possibly a new access network. The selection of an optimal access point also depends on the user context, for example the desired user services or cost restrictions. In the following, we discuss typical context information that is related to handover decisions and present the problem of providing this information to the mobile node. In general, context information comes from different sites, layers and device entities. Examples are layer 1 and 2 information regarding signal and link information, as well as application-layer information both on the mobile device and on the network side.

Table 10-1 gives a classification of typical context information. The table is clearly just a snapshot, as new types of context information may appear. For instance, in future ad hoc networks, new context data such as user groups may be relevant. Some items such as the user profile may appear multiple times, as the information is spread over the user device, the operator and possibly separate service providers.

As an example, the user profile may include subscribed services and service preferences (such as which services have to be downgraded or dropped if insufficient resources are available).

Table 10-1. Context information classification

	Context information on mobile device	Context information on network side
Static	User settings & profiles Application settings	User profile and history Network location Network capabilities and services Charging models
Static w.r.t. current cell	Reachable access points	Potential next access points
Dynamic	Application requests Device status (battery, interface status, etc.)	Location information and location prediction Network status Network load

The problem is that potentially all of this context information is relevant, but using it is difficult for several reasons:

- There is very limited time for many network layer tasks, where decisions have to be taken fast. In addition, there may be no or poor network connection at the time of the decision. Hence, context information should be collected and provided before this point in time.
- Context information is distributed and not readily available at a single network entity. For instance, some context information may be available in the home network of a user, some may be available in the visiting network and some may reside on the node.
- Dynamic context information may change frequently or lose accuracy over time. For instance, it is tempting to transfer information about the current access point load, yet its relevance decreases quickly over time.
- The relevant context information and the methods to interpret the context information may evolve over time. Hence new algorithms for interpreting the context data are needed. For instance, consider the following cases:
  - where the user profile changes (for example new services are added, which need to be considered for choosing an access point)
  - roaming agreements regarding services change
  - if the location prediction method is improved or changed (for example for cars or trains)

Consider, for instance, the context-aware handover problem. The main idea is that using the context information can avoid handover disruptions and wrong handover decisions, which degrade the user experience. Clearly, some of this information contains uncertainty and hence the choice is not guaranteed to be the best in all cases. Also, the transmission of the information to the mobile device and the processing of the information creates some overhead.

In most cases, applications only use a small fraction of the available context information for specific services. In the particular case of mobile networks, there are some further problems, such as unstable and resource-scarce wireless links, power limitations and intermittent communication while roaming. In addition, heterogeneous devices and communication systems, as well as client–provider relationships, have to be taken into consideration.

10.2.1 Related Work on Context Management and Context-aware Handover

More intelligent handover procedures have been considered by Stemm and Katz[8] and Pahlavan *et al.*[9] However, the parameters that are taken into account for the handover decision are still confined to the type of the radio access technology as well as to the signal strength. Stemm and Katz used different handover policies for heterogeneous networks, considering as handover parameters mainly the air-interface type and the available bandwidth at the access router. Chan *et al.*[10] used fuzzy logic to cope with the complexity of the information set retrieved from a heterogeneous mobile access network. However, the focus is on the algorithm used for the handover, and how to apply fuzzy logics to this problem. Another related paper is that by Kounavis *et al.*,[11] who provide a framework for a programmable handover, which covers mobile-controlled, network-controlled, network-assisted and mobile-assisted handover. Our work can extend this in order to adapt to context information. To summarize, the baseline of these papers is that a more intelligent handover is needed for a heterogeneous access network. However, none of these papers presents a general framework for a handover mechanism that benefits from the various contexts with which the mobile node might be confronted.

10.3 Context Management Approaches for Context-aware Handover

In this section, different approaches to context-aware handover will be presented and we will see in which way the diverse context information can be processed by the network and the mobile device. This example will serve as the main example for developing a more general framework in the section below. As the context information in the network may be spread out over several entities, such as the location information server, we consider one network element where the context information for one user is collected proactively in order to prepare the handover. This section is an extended version of that given by Prehofer *et al.*[12]

In our approach, the following main entities are defined, as shown in Figure 10-3:

- handover decision point, which decides on the access point to be taken for handover, i.e. the mobile node in Figure 10-3
- context collection point in the network side, which collects and compiles the relevant context information from different sources; this is then delivered to the handover decision point

We assume that the handover decision takes place inside the mobile node, but is possibly controlled by the network with the supplied context information and rules. This is important, because many pieces of dynamic context information are only available on the node (for example the signal strength). The level of control from the network side just depends on the logic implemented in the mobile node and how it reacts to network context information. This means that we view a network-controlled handover as a special case.

The problem that we address in the following is hence to supply the handover decision point with the appropriate context data in a timely and effective way. Furthermore, the logic to interpret the context data must be agreed on beforehand or it must be supplied as well. We have to balance between several conflicting requirements. On the one hand, it would be useful to supply as much relevant information as possible and to update this information in the case of change. The algorithms or logic in the nodes should also be flexible and generic in order

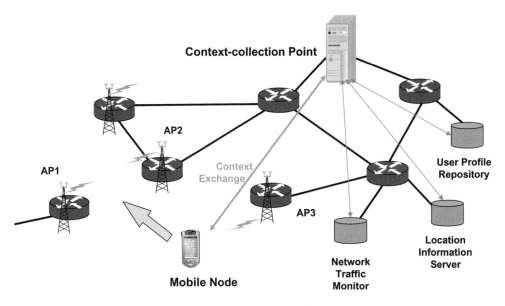

Figure 10-3. Architecture for context-aware handover

to adapt to new requirements (for example to handle new pieces of context information). On the other hand, the system has to be scalable and should minimize overhead, in terms of both communication cost and computational resources.

In the following, we present two approaches for handover decisions, which we then combine to a flexible framework.

10.3.1 Context-exchange Protocol Approach

The classical approach for context exchange is to define a context information exchange format and attributes (for example using appropriate XML dialects such as CC/PP[13]). In addition, a protocol for exchanging context information between the context collection point and the mobile node is needed. This means that interpretation of the context information should be fixed and possibly standardized. The appropriate algorithms need to be implemented in the nodes. The algorithms to collect the context information on the network side can be specific to the network or the location.

In summary, this approach requires the following pieces, which must be deployed in nodes before the procedure can take place:

- context format and attributes agreed between the network and the nodes
- an algorithm for context interpretation and handover decision on the nodes
- a context collection mechanism that compiles the relevant information in the specified format on the network side
- a protocol to exchange context information, preferably with the option to provide selective updates of information

In addition to the algorithms, triggers that initiate the collection and transfer of the context information at the right time are needed. This again can be context driven. For instance, the context information that a node is moving out of the coverage area of a cell may trigger the collection of other context.

To perform a context-aware handover, the following steps are performed by the handover decision and context collection points:

1 Exchange context data before the handover decision is imminent (for example while the user is still close to the old access point).
   a If needed, updates of this information may be sent repeatedly.
2 The mobile device uses the following handover procedure, which is triggered by loss of connectivity or by the need for different access point:
   a Collect the dynamic context of the mobile device (visible access points, signal strengths, application status/requests).
   b Invoke the algorithm with the provided data and the dynamic context to decide on the appropriate handover access point.
   c Execute the handover to the chosen access point.

As an example, we refer to the scenario above in Figure 10-1. Let us assume the context format is simply a preference list of APs with QoS capabilities. The context information is compiled on the network side. The resulting preference list optimizes the user service based on the user/network context available at the context collection point. The capabilities of the APs are shown in 10-2. Let us assume that the user moves with a car and the typical preference is voice. The context collection point first gathers this information and compiles it as follows:

1 Based on movement prediction, the preferred APs are: AP2, AP4.
2 We order this based on preferred services: AP2, AP4.

This list (AP2, AP4) is sent to the mobile device before handover. During handover, the algorithm on the device will use the dynamic information about reachable APs and then take the highest one of these in the list. In this example, the control functionality is mainly on the network side. An advantage is clearly that the data volume sent to the node is small.

A variation with more responsibility on the node side is the following. Assume that the transferred context is extended to include the properties of the above list for each AP. Also, the decision algorithm considers a more dynamic context (application requests, ongoing

Table 10-2. Example access point capability table

	Air interface	Operator	QoS 1 (voice)	QoS 2 (video)	QoS 3 (data)
AP 1	UMTS	A	+	+	+
AP 2	UMTS	B	+	+	+
AP 3	GSM	B	+	−	−
AP 4	WLAN	A	−	+	++

sessions with respect to the QoS classes in the above table). In this case, the context collection point makes some pre-selection and ordering of APs and sends the table with these APs to the mobile device. The algorithm in the device applies the dynamic context to select the suitable entries that are available and match the current session/application requests. The algorithm always checks the highest entry in the table (preference ordering by collection point) at first.

In both approaches, the information sent to a node is optimized for this device, which saves transmission and processing cost for each individual device. On the other hand, some context information may be relevant to more than one node and can be sent via broadcast to all devices attached to an access point. Additional information, which is not relevant for all nodes, can be sent as updates to specific nodes. In this way, the processing at the nodes is more involved, as they have to filter and process more information.

A limitation of both approaches is that the processing of context in the node is fixed. In the case where new context information (for example new attributes or finer-grained information) is available, the network has to translate this to the agreed format and attributes. As the lifetime of a mobile device is several years, this is quite a limitation. This limitation will be addressed in the following approach.

10.3.2 Dynamic Download of a Context-aware Decision Agent

In this second approach, we do not send context information in explicit form to the mobile node. Instead, we download a software agent that encapsulates both the context information and the algorithms for interpretation of the data. In this way, we can avoid the limitations of a fixed context-exchange protocol. Regarding flexibility, this approach is clearly superior to the first one and also does not require the standardization of context formats and their interpretation. We assume for this solution that the handover decision point is programmable to support the dynamic installation of an agent (for example using active network technology[14]).

This approach consists of the following steps:

1   Prepare an agent at the context collection point with the algorithm and the collected context data needed at the handover decision point. This takes place when the context or algorithm changes significantly, for example when entering a new cell or upon changes in the user profile.
2   Download the agent proactively before the handover to the handover decision point.
3   The agent is invoked at the time of the handover as follows:
    • input: dynamic node context (for example reachable access points, application requests and sessions)
    • output: handover decision (i.e. the access point has to be chosen)

This procedure is illustrated in the sequence diagram of Figure 10-4. The preparation of an agent in the first step can be implemented in different ways. One option is to select one from an existing library according to the situation. Alternatively, software pieces can be parameterized or composed from a toolbox according to specific rules. The main advantages of this approach are flexibility and fewer assumptions about the mobile nodes. We assume only that a software agent can be installed, for example on a mobile Java platform. This agent must

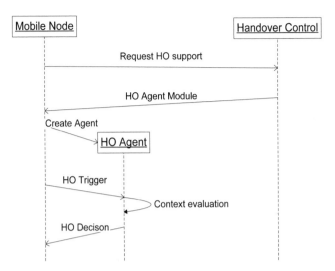

Figure 10-4.   Sequence diagram for agent-based approach for context-aware handover

have interfaces to access the dynamic context information and to control the handover. In
this sense, the local interfaces of the agent are now the main limiting factor.

Coming back to the handover example, we assume a user who is moving in a car and
whose preference is low-cost voice and occasional fast data service. With this context information, the handover collection point devises an agent that decides for AP4 only if the data
service is selected. The decision algorithm of the agent that is downloaded from the network
computes the preferred access point is as follows:

Procedure CalculatePreferredAP
Dynamic Local Input: RequestedUserService
If (RequestedUserService = Data)
   then preferredAP = AP 4
   else preferredAP = AP 3

Note that the agent is invoked at the time of handover and hence uses dynamic context information from the mobile (for example current user services). Compared to the above approach,
only the agent has to be downloaded, not a complete table. The agent can be seen as a simple
script, which contains a decision tree compiled from the essential context data. Depending
on the run-time environment and language, such an agent may just be slightly bigger than a
table with context information.

Another, more involved example is the following agent:

Procedure CalculatePreferredAP
Dynamic Local Input: RequestedUserService, RequestedDataRate
if (RequestedUserService = Data)
   then if RequestedDataRate < 50 KBit/s)
      then preferredAP = AP 3
      else preferredAP = AP 4

else
If (RequestedUserService = Video AND RequestedDataRate > 15 KBit/s)
   then preferredAP = AP 2
   else preferredAP = AP 3

Here the transferred agent is slightly bigger than in the example above, as it contains the decision logic. In general, the agent can of course be considerably larger and may contain more explicit context data and algorithms. This is also an advantage of such an agent – the complexity and amount of data can be adapted to the user and the user situation. Similar to the above, it is possible to broadcast the agent to several nodes at the same time to save resources.

An issue not considered here is the security of the agent solution, as we have to avoid malicious or wrong code. For this, we assume that the agent may be executed in a secure software execution environment. Furthermore, termination of the agent (possibly by the execution environment) must be ensured. A potential drawback of this solution is the overhead for downloading the code; yet in many cases the total amount of data to be transferred is the same as or less than in the first approach, as only the appropriate code needs to be down-loaded. Furthermore, we can rely on standard routines, which do not have to be downloaded each time.

10.3.3 Combining the Two Approaches

The above approaches are two extreme solutions to the problem: in the first one, static format specifications for the context information have to be agreed upon between the network and all the mobile devices; in the second one, an agent has to be downloaded from the network by the mobile device in order to interpret the context information and to process it. The difference is that here the agent solution is completely flexible with respect to the data structures and formats used for the context information. In contrast, the first solution is limited to a fixed data structure (for example a table or a decision tree).

In the following, we combine the two ideas. The algorithm and the context information are downloaded separately on different time scales. We use an agent that is also capable of context exchange, and exchange the agent only if there are significant changes in the processing or structural changes of the context information. An example of the latter is a change in the context format, user profile or context processing algorithm, for example for a new charging model.

The overall process consists of the following two procedures, which are used independently on different time scales:

- The steps to exchange the agent (decision algorithm) are:
  1 Prepare an agent at the context collection point, including the algorithm and the collected context data needed at the handover decision point.
  2 Download to the handover decision point the new agent, including the algorithm and the context exchange protocol.
- The steps for context exchange and handover are:
  1 Exchange context data if the context changes, typically when entering a new cell.

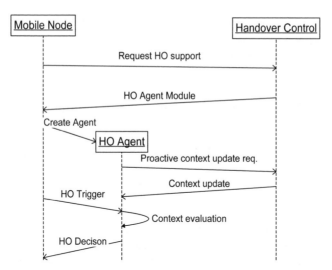

Figure 10-5. Sequence diagram for agent-based approach with context exchange

2 The agent is invoked at the handover time as follows:
 • input: dynamic node context and transferred context
 • output: access point to be chosen

This procedure is illustrated in the sequence diagram of Figure 10-5.

As an example of this scheme, we reuse the previous example. Assume the agent is as shown in the pseudocode below, using four parameters APL1, ..., APL4, which are lists of access points that group the APs in the vicinity by capabilities. We assume a process ExchangeContext, which is permanently active and waits for updates from the context collection point. The main procedure calculates a list of preferred access points for handover. If the list is empty, then any access point can be chosen.

Process ExchangeContext
  Repeat
    Wait_event from ContextCollectionOoint
    Receive APL1, APL2, APL3, APL4 from ContextCollectionOoint

Procedure CalculatePreferredAPList
Dynamic Local Input: RequestedUserService, RequestedDataRate
if (RequestedUserService = Data)
  then if RequestedDataRate < 50 KBit)
    then preferredAPlist = APL1
    else preferredAPlist = APL2
  else
  If (RequestedUserService = Video AND RequestedDataRate > 15 KBit)
    then preferredAPlist = APL3
    else preferredAPlist = APL4

Note that an agent update can be applied on different time scales. For instance, each time the user context changes, for example when the user enters a car, a new agent is installed. Alternatively, the agent can be updated only if there are changes in the user or network services, which may happen just a few times a year.

10.4 Framework for Context Management in Mobile Networks

Based on the discussion of context-aware handover in the last section, we now aim for a more general framework for context management, which is suitable for different network functions. A first version of this framework has been presented by Mendes *et al.*[15] Such a framework should be able to collect, process and distribute non-structured and short-lived information, which can be located in distinct network nodes and devices. This framework should also ensure that the right information is available at the right time to a wide range of applications. An important issue in the design of such a context management framework is the consideration of different application requirements. The requirements of applications can differ, for instance in terms of the information exchange patterns, of the data model of context information and of the granularity and amount of the transmitted data. The last of these is particularly important in mobile networks, since scarce wireless resources limit the amount of information that can be exchanged with wireless devices. Another important requirement is the capability to use cross-layer interfaces efficiently to collect information in different protocol layers.

In the following, we introduce a general-purpose framework able to manage dynamic and distributed context not only at the application layer, but also at the network and link layers, while clearly separating application-specific context from general context. The division between specific and generic contexts allows the creation of generic mechanisms to collect context information. The adjustment of generic context information to the needs of applications and the network service is done by specific adapters.

The main goal of the proposed framework is to facilitate the collection, pre-processing, distribution and use of distinct context information by a wide range of applications. The first obstacle to creating a general-purpose flexible framework is the application-specific nature of the context information to be collected and the distribution mechanisms to be used. To solve this problem, the proposed context management framework is divided in two distinct parts. One subset of modules is generic to any type of application, while the other is application specific, as illustrated in Figure 10-6.

Note that we focus on the application-specific adaptations in this section. We do not detail the application-independent context exchange. This requires generic modelling of context information and efficient, scalable methods for processing context information.

Since each application has its own characteristics, the generic context information has to be translated to a format that each application can understand and its distribution has to be done by means of a protocol that is suitable for each application. Therefore, in the proposed framework, the selection of context information is made available by a *Generic Context Collection Point* (GCCP). One GCCP is able to establish an association with different context suppliers and with other GCCPs, as shown in Figure 10-6. The adjustment of the generic information to each application is done by an application-specific adapter, called a *Context Service Adapter* (CSA).

The collected generic information is distributed among separate applications. However, each application, which is called a *Context Client* (CC) in the proposed framework, has its

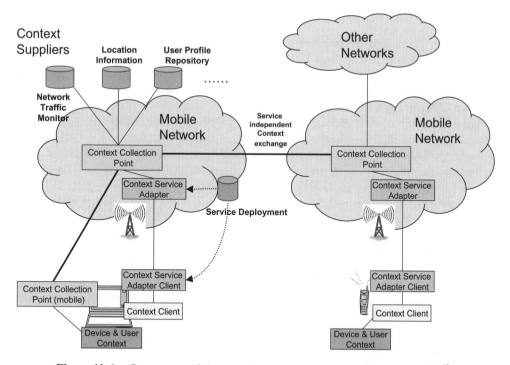

Figure 10-6. Components of the context management framework (from Mendes[15])

own specific requirements. For instance, different applications can request information with distinct data models, such as hierarchical or flat data, as well as request the transmission of different amounts of data with different granularities by using distinct exchange patterns, such as client–server, peer-to-peer or reactive–proactive.

A set of distinct CSAs may interact with the same GCCP, since the latter can supply generic context information to a wide variety of applications. Each CSA sends the adapted information to one *Context Service Adapter Client* (CSAC), which interacts with a specific CC. The inclusion of these two elements in the application-specific context management, between the generic GCCP and the CC, allows a more flexible distribution and utilization of context information.

Since different sets of CSA and CSAC modules can be required at different times and places, the proposed context management framework includes a *Service Deployment Framework* (SDF), capable of controlling the provision of application-specific services in a network.

In the specific case of mobile networks, special attention has be paid to the definition of application-specific modules, because of the scarcity of wireless links and mobile devices. The reduced capacity and potential instability of wireless links require low data-transfer rates between the network and mobile devices. Furthermore, the limited energy resources of mobile devices require most of the computation to be kept on the network side. Section 10.5 illustrates the application of the proposed general-purpose context management framework to mobile networks.

10.4.1 Architecture to Manage Context Information in Mobile Networks

This section presents an architecture that implements the proposed context management framework to deploy and manage different types of services in mobile networks. In this architecture, application-specific adapters are installed on demand by a service deployment infrastructure. Figure 10-7 illustrates an architecture in a scenario with two mobile networks and two different types of mobile device.

The exchange of context information among GCCPs is application independent. However, the configuration of the exchange protocol depends on the location of the GCCPs. For instance, policy issues can restrict communication between GCCPs located in networks belonging to different operators, while communication with GCCPs located in mobile devices should be moderated to avoid wasting the scarce wireless resources.

To cope with the limited resources of wireless links and devices, the proposed architecture concentrates most of the application-specific computation into CSAs located in the network, while the CSAC module is located in mobile devices.

Context information needed by mobile devices can be requested not only from a local GCCP or a GCCP in the network to which the mobile device is attached, but also from GCCPs located in other networks. Retrieving context information from other networks depends upon the policies that rule the protocol between operators. For example, different operators may agree to exchange information about their network capabilities, but not about their client's preferences.

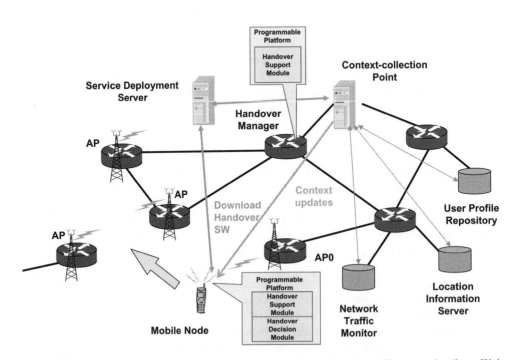

Figure 10-7. Network architecture to manage context information in mobile networks (from Wei *et al.*[1])

Although CSAs and CSACs are application specific, they may also be specific to a network operator. This means that after roaming to another network, mobile devices may or may not keep the CSAC that they were using on their home network. In either case, the communication between the CSAC in a foreign network and the CSA in the home network is only limited by the roaming policies between operators.

However, it is assumed that a network provider can only deploy CSAs in its own network, and CSACs in mobile devices that request them. The implementation of a SDF requires a *Service Deployment Server* (SDS) in the network, as illustrated by Figure 10-7, and this stores the description of all available services in the operator's network and controls the deployment of such services.

Application-specific services can only be deployed in programmable nodes, i.e. network elements that can install and use different modules. This capability is provided by a *Service Creation Engine* (SCE), which is able to request services from a SDS, and by programmable platforms.[16] These allow the implementation of application-specific modules in network nodes and mobile devices as a result of its flexible programmability characteristics. Note that one device can have more than one programmable platform for different purposes.[16]

10.5 An Implementation of Context-aware Mobility Management

For an implementation of the above framework, we now focus on how to combine the different components. For a realistic environment, we have to address several issues. First, the deployment of the software modules has to be managed in a scalable way. Second, we need to select an execution environment for the mobile node. The deployment and execution in the mobile nodes has to be resource efficient and must also consider the performance limitations of the devices. Furthermore, the process should be seamless on the mobile device, not interrupting the service.

In the following, we integrate the context management framework with an active platform and a service deployment scheme to provide the functionalities needed for context-aware handover (HO). The context management framework is in charge of collecting the relevant context information for different services and managing the context information. The active platform is used to exchange and process the context information. The service deployment scheme is used to synchronize and manage the working of the active nodes involved.

In our network architecture, shown in Figure 10-7, context information is stored in context information repositories, such as a Location Information Server (LIS), a Network Traffic Monitor (NTM) and a user profile repository. The LIS is responsible for tracking the position of each mobile device in the provider's network and has knowledge about nearby APs, while the user profile repository stores the user profiles as seen by the network service providers. The NTM is used to monitor the available bandwidth of different APs. Moreover, we introduce a Handover Manager (HM), which controls the handovers carried out in some part of an access network. The HM is responsible for filtering and processing HO-related context information. Finally, a Service Deployment Server (SDS) is used to manage and install the service modules needed in the network nodes and the mobile nodes.

The mobile nodes include two modules, called Handover Support Module (HSM) and Handover Decision Module (HDM), which are client service adapters. The handover manager on the network side also includes a Handover Support Module, which is a context service adaptor.

Implementation based on Active Networks

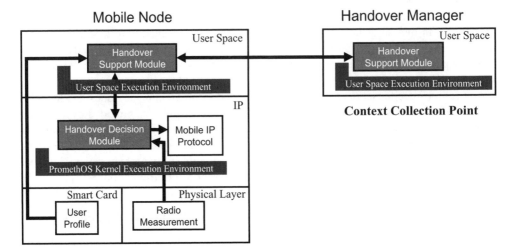

Figure 10-8. Programmable software modules for context-aware handover (from Wei *et al.*[1])

Figure 10.8 shows the execution platforms of the nodes, which are the execution environments of the active modules. For example, to realize a context-aware HO service, we need to install a context exchange protocol (HSM) and a handover decision mechanism (HDM) on the related nodes, i.e. on the active platform of the mobile network (MN) and HM. This is realized by service deployment.

The HO decision algorithm is implemented as an HDM, which is installed in the MN, and the context exchange protocol is implemented as an HSM, which is installed in both the MN and the HM. The HO decision point uses simple algorithms (such as a rule-based logic) for interpreting the data delivered by the context management framework. It is the task of service deployment to ensure that both the HO decision point and the context collection point are proactively supplied with the appropriate algorithms.

10.5.1 Active Node Platform and Service Deployment

In the following, we describe the node platform for our architecture. We use active networking technology to meet the requirements outlined above. Since the algorithms executed to determine the best AP for handover are context-dependent, the network elements and the mobile end systems involved in this process need to be programmable. Active networking technology is a good candidate to fulfil this requirement. Our active node consists of the basic processing hardware, a node operating system and several execution environments, in which active applications will execute the handover algorithms. The node has to support the dynamic installation of a handover decision module at run time, without interrupting the node's proper working. The active node architecture and implementation we chose for our system is

PromethOS.[17] PromethOS is a generic platform for running active applications in a Linux environment, allowing for on-demand installation of user- or kernel-space modules. This flexibility, as well as the generality offered by the Linux environment, motivated us to choose this platform.

While PromethOS provides a basic active-node platform, our application scenario also needs to support the selection, installation, configuration and management of the service components, which are discussed in the next section.

10.5.2 Node-level Service Deployment

Node-level service deployment comprises selecting, downloading, installing and configuring implementations of service components on an active node, such that these components jointly provide the specified service. The Chameleon service deployment framework is well suited to accomplish these functions for node level service deployment.[18] Chameleon uses a service specification – generated by the network-level service deployment – and a description of the intrinsic properties of the active node to determine which implementations of service components need to be installed on the active node. The service specification is given as an XML document, which conforms to an XML document type definition describing the structure and format of service specifications. Chameleon resolves the service specification against the description of the node properties, thereby creating a tree-like structure representing all possible implementations of a service. In our application of Chameleon, such service specifications are generated by our network-level service deployment scheme, supporting context-aware handovers.

10.5.3 Network-level Service Deployment Scheme for Context-aware Handovers

Network-level service deployment is in charge of providing the software to the node from the network side. Service deployment can be of two kinds, provider-initiated or user-initiated. In the former case the service is deployed over the network in advance, long before the arrival of any service user. In the latter case the deployment is on demand and the arrival of the first service user initiates the installation of the service in the network.

We use a simple, centralized network-level service deployment scheme. Its core is a central management entity, called Service Deployment Server (SDS), as illustrated in Figure 10-7. The SDS contains a Service Deployment Manager (SDM) module that controls the network-wide signalling and all related synchronization functions needed during service deployment. Moreover, it contains a Service Server, which stores the descriptors of the services known to the system, and a Code Server, which stores the implementations (code modules) of the service components available in the provider's network. These servers are managed by the SDM and they can be located anywhere in the network.

10.5.4 Putting Everything Together

The full operation of our context-aware handover testbed is explained in the message sequence diagram shown in Figure 10-9. Step 1 is the service deployment, which includes fetching the right service components for a required service, installing them on the appropriate network node and confirming the successful installation of all the components for this service. Step

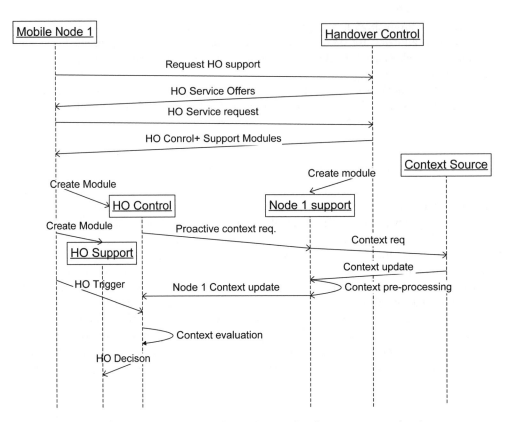

Figure 10-9. Sequence of signalling and processing for context-aware handovers

2 collects relevant context information. In Step 3, the context information is evaluated and the HO decision is made.

After the HDM makes the decision on the target AP, based on the collected context information, the decision is sent to the mobility management component of the mobile node to execute the handover. In our approach, mobility management is done with Mobile IP.

10.5.5 Evaluation Scenario

This section gives an overview of the evaluation of the proposed architecture for managing handovers with the goal of optimizing the selection of the new access network. The evaluation is done by using a prototype that reproduces the environment illustrated in Figure 10-1.

In this evaluation process, two types of context information are considered: the user's location, speed and trajectory, which depend on how the user moves (walking/inside a train); and the Quality of Service (QoS) required by the application. If the user's location context is being considered, the handover can be made to a network that has a better coverage along a railway line, or to a network with small range but better quality if the user is walking. If the QoS required by applications is being considered, an additional criterion can be the network traffic, which can affect the quality level perceived by the user.

10.5.6 Experimentation and Evaluation

The customized handover service is evaluated on the basis of a scenario in which the user is located in a train and starts to watch a video streamed by a corresponding node.

Experimental results show that with context information, mobile devices have greater control over their handovers. The results also show the importance of using the right CSA and CSAC modules within different contexts.

When the mobile device moves in the direction of AP3 and FAP4, the SCE gets information about the user's location from the GCCP installed in the mobile device. Since the user is on a train, the SCE on the mobile device installs the two components of the CSAC module in the user space and kernel programmable platforms, as shown in Figure 10-8.

The corresponding CSA module is installed in the programmable network node by the local SCE. The new CSA processes the information collected by the GCCP installed in the network node about the coverage of AP3 and AP4 near the railway line. The new CSAC processes the information received from the CSA and sends it to the CC. The installed customized modules allow the mobile device to register with AP1, since this is the network with better coverage near the railway line.

To continue this experiment, we add extensions that consider the application demands in the mobile node, here video. For this purpose, new CSAC and CSA modules are installed to handle information related not only to the user's location but also to the load on the access networks. Based on this context information, the CC may decide to hand over to AP4, because this network has a smaller traffic load than AP1. As a result of this handover decision, a good video quality is kept.

Some measurements were made to evaluate the impact that the proposed architecture has on the network and on the applications,[1] namely the signalling overhead, the time required to deploy context-aware services and the time required to collect customized handover information.

In the current prototype, the service deployment in the programmable network node takes less than one second, while it takes a few seconds in the mobile device. The gathering of context information takes between one and two milliseconds, from the moment the CSAC requests handover information until the moment it receives it. This time corresponds to the round-trip time between the mobile device and the network programmable node, since in the current prototype the programmable node that includes the GCCP also emulates the functionality of the location information server. Since the GCCP pro-actively collects context information and the time required to retrieve this information is very small, the probability that handovers need to be made without context information, i.e. using only the MIPv6 functionality, is very low.

10.6 Summary and Concluding Remarks

The main contribution of our approach is the definition of a general-purpose framework able to manage dynamic and distributed context, not only in the application layer, but also in the network and link layers, while clearly separating the application-specific context from the general context. This division allows the creation of generic mechanisms, which are transparent to applications and network services, for collecting and processing context information. The adjustment of the context information to the needs of each application or network service is made by specific adapters. In the architecture that implements the proposed context management framework, these application-specific adapters are installed by a service deployment infrastructure. Adapters on the network side can be installed on demand or in advance, depending on the configuration of the service deployment infrastructure. Adapters required on mobile devices are always installed on demand.

Experimental results, based on a prototype implementation, illustrate the capacity of our architecture to handle different types of context information and context adapters. Our approach can enhance the performance of network services, such as handovers, without causing any deterioration in the performance of the mobile networks. Furthermore, the use of customized modules, while creating some initial deployment overhead, is shown to be very efficient.

Other recent studies have proposed more sophisticated procedures for network services, by increasing the services' awareness of the surrounding environment. However, most of these approaches focus on the study of handover,[19] considering a limited number of context parameters, such as the type of the radio access technology and the signal strength, or they consider the heterogeneity of access networks.[8] Few projects address the flexible collection, processing and distribution of context information, as has been done in the Context Toolkit[20] and the Solar project.[21]

The author would like to thank Qing Wei, Paulo Mendes, Nima Nafisi, Bernhard Plattner and Károly Farkas for discussions and contributions to the project.

11

Contextual Intelligence

Matthias Wagner, Marko Luther and Massimo Paolucci

11.1 Introduction

Performing complex tasks over the Web has become an integral part of our everyday life. The advent of mobile services will add to the broad range of existing services offered on the Web and provide additional features such as location-based information. The success of these future mobile services will largely depend on their ability to maximize their value in a varying context. Contextual intelligence in devices, mobile applications and service platforms is needed to manage different mobile terminals, to personalize content and services or to narrow down possibly very large sets of applicable services in a given situation. By 'contextual intelligence' we mean added capabilities of mobile services and applications that allow flexibly adapting to varying contexts and maximizing their value therein. To this end, we exploit Semantic Web technology as an enabler for this added intelligence in context-aware mobile services. Our vision is to overlay the Semantic Web on ubiquitous computing environments making it possible to represent and interlink content and services as well as users, devices, their capabilities and the functionality they offer. With this vision of Contextual Intelligence in mind, we exploit the use of OWL and related technologies with a particular focus on providing access to the Semantic Web for mobile applications.

The aim of this chapter is to provide a brief and concise overview of our project activities connected to Contextual Intelligence. As an additional initial motivation, we present our notion of personalized user-centred Web services provisioning. Here, we focus on the different phases of interaction with services and show how cooperative discovery algorithms can essentially improve service provisioning.

11.2 Research Prototypes

Figure 11-1 shows the range of projects related to the Semantic Web that we are currently pursuing in our laboratories. On a research map, the ongoing activities can be aligned

Towards 4G Technologies. Edited by Hendrik Berndt.
© 2008 John Wiley & Sons, Ltd.

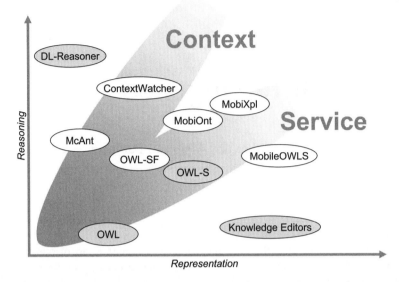

Figure 11-1. Project activities concerned with Semantic Computing and the Semantic Web

according to their level of concern with OWL fundamentals in terms of representation capabilities and reasoning support. The projects can also be projected onto an additional application dimension representing activities that target semantic-based mobile applications in terms of support for context representation and management, support for semantic mobile services or both of these aspects. We shortly present our project activities in the following.

In particular, the practical success of Semantic Computing and the Semantic Web will largely depend on whether W3C's recommendation for the Web Ontology Language OWL can receive broad acceptance and be applied to a critical mass of industry-strength applications. We have therefore been exploiting the use of OWL with a particular focus on tool support and on providing access to the Semantic Web for mobile applications. In relation to the latter our vision is to overlay the Semantic Web onto ubiquitous computing environments, making it possible to represent and interlink content and services as well as users, devices, their capabilities and the functionality they offer. In this paper we present our first experiences and lessons learned from early work and try to give constructive feedback for possible enhancements of OWL and its tools.

11.2.1 OWL-SF

Major challenges in designing ubiquitous context-aware systems, such as the distributed nature of context information and the heterogeneity of devices that provide services and deliver context, have been approached by developing a distributed semantic-based service framework called OWL-SF.[1] In that framework, W3C's ontology language OWL[2] is used to capture high-level context elements in semantically well founded ways. Devices, sensors and other environmental entities are encapsulated using OMG's Super Distributed Objects technology (SDO)[3] and communicate using the Representational State Transfer model (REST).[4] The framework was evaluated by implementing an early case-study that uses enhanced presence control to realize intelligent call forwarding.[5]

11.2.1.1 Functional Architecture

The functional architecture of the OWL-SF framework integrates two basic building blocks, namely OWL-SDOs and Deduction Servers (DSs) (Figure 11-2). A system may be composed of multiple components of both types that can be added and removed dynamically at run time. Deduction Servers provide reasoning support based on the ontology structures collected from the accessible OWL-SDOs and perform service calls on the OWL-SDOs based on the derived conclusions. In contrast, OWL-SDOs implement a certain program logic that encapsulates the underlying hardware and software components providing an OWL-enhanced SDO interface. In general, components of both types could provide the same interface (i.e. the DS could support the OWL-SDO interface or an OWL-SDO might contain a local reasoner). However, as the intended functionalities of the two are conceptually different, they are treated separately in the architecture description.

Two kinds of reference points are used in this architecture. One, the SDO reference point, denotes the communication among OWL-SDOs and between DSs and OWL-SDOs, while the other, named SDO*, establishes the link among DSs. Service calls are realized and status information on the internal state of SDOs is exchanged via the SDO reference point. In contrast, the SDO* reference point is used to exchange knowledge between separate reasoning servers. It provides the same interface as the SDO reference point, although the semantics of the exchanged information differs.

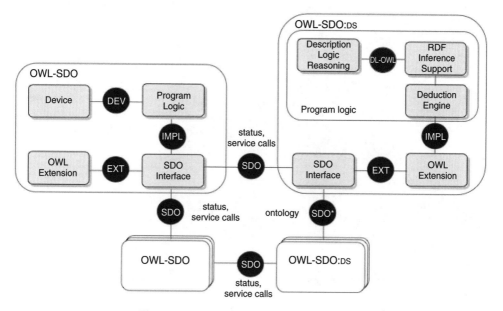

Figure 11-2.   OWL-SF functional architecture

11.2.1.2 Context-aware Presence Management

Based on OWL-SF, a scenario for intelligent call forwarding was realized that targets the office environment and demonstrates context-aware presence management to enable intelligent call forwarding. The scene is populated with several persons holding mobile phones who are either in an office or in a meeting room. Depending on the location of a person, the people in proximity and the current time, the person's situation is classified and further used to determine if that person will receive phone calls or if she is currently busy and phone calls should be redirected.

The demonstrator (Figure 11-3) is written in Java and features a rich GUI that visualizes people, mobile phones, door panels and rooms. The user can interactively modify people's contexts, which changes people's situations and thus the call forwarding. Mobile phones that are connected to the demonstrator via Bluetooth or Parlay are configured accordingly. Additionally, further devices can be influenced, based on people's situation, by specifying rules. For example, the jukebox in the meeting room can be controlled by rules that can be modified by the user in the same way as any other rule in the system, via a rule broker component. For simplification purposes, time changes in the simulation are handled by allowing the interactive setting of one of the two abstract states, office hour and lunch hour. Similarly, the process of the location detection of persons is simulated by using memory sticks that are mounted or unmounted on certain ports on a computer.

Figure 11-3. OWL-SF simulation environment

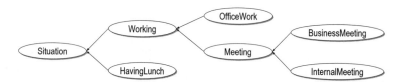

Figure 11-4. Fragment of the situation ontology

The actual situation of an actor and their availability for phone calls is decided by using concept classification. Figure 11-4 shows a fragment of the corresponding OWL concept hierarchy. For example, an individual person's situation is classified as Working when it is a Situation and the property has CurrentTime relates to an individual of type OfficeHour. Accordingly, a person is classified as being in a Meeting situation if he is Working, his location is classified as MeetingRoom and he is together with other persons.

The OWL-SF architecture has been influenced by the Context Toolkit (CT),[6] which structures contextual environments into Context Widgets that encapsulate context sources, Interpreters that raise the level of abstraction of pieces of context and Aggregators that collect multiple pieces of context together. However, the CT only provides weak support for knowledge sharing and high-level context reasoning because context information lacks formal semantic descriptions.

Other approaches such as CoBra[7] and CALI[8] tackle these issues by using OWL descriptions, but do not define a fully distributed architecture. Both approaches assume one central reasoning engine. CoBra relies solely on rule reasoning, which cannot be complete for OWL (and not even for OWL-Lite[9]). CALI tries to overcome the limitations of DL-based reasoning using a hybrid reasoning approach that tightly integrates DL and generic rule reasoning.[10] However, such a tight combination might easily lead to an inability to decide because rules can be used to simulate role value maps.[11] Furthermore, the initial results of our own experiments indicate that an ad-hoc approach to hybrid reasoning easily becomes computationally expensive.

Our early study demonstrated that combining the SDO technology with OWL to create a working fully distributed semantic service environment is feasible. However, we need to extend our approach towards the integration of privacy and low-level context reasoning mechanisms. The latter will especially raise new challenges as context gathered by sensor devices is hardly precise and therefore support for uncertain representations has to be integrated.

11.2.2 ContextWatcher

Within the IST project MobiLife we have implemented ContextWatcher (Figure 11-5), an early prototype for semantic-based monitoring of mobile users. The project aims at frameworks for context-aware services that support users in their daily life. OWL upper context ontologies define the basic contextual categories and the relations among them. Such high-level structuring of context information enables its integration and consolidation on a semantic basis. Furthermore, the axiomatic descriptions of context elements such as personal

Figure 11-5.   ContextWatcher mobile client

situations (i.e. Working, At_home, etc.) can be directly used by logical inference engines to realize reasoning about the user's presence and virtual location.

11.2.2.1 Context Sharing

ContextWatcher make it easy and unobtrusive for people to share information about their whereabouts, or any other piece of context information they might want to share, ranging from body data to pictures, local weather information and even subjective data such as moods and experiences. The goal of the application is to offer people new ways to keep (and stay) in touch without the need for direct interaction. One example is to browse a friend's on-line photo book every week; the contextual descriptions with the photos give a good impression of his activities, providing food for talk next time you meet him. Examples of context sharing include:

- *Context-augmented buddy lists.* This is real-time sharing of context information between two persons. Before sharing, two persons have to agree to become buddies and share specific types of context information. At any moment in time, buddies can see, for example, where there friends are, which books they read, what the weather is in their friend's place, how their friends feeling and whom they are with. This information is presented in the form of a buddy list that is common for most instant messaging applications, but in ContextWatcher it also provides information about the whereabouts and other detailed information in the context menus, see Figure 11-5. Additionally, buddies can be grouped in an ad-hoc fashion based on context information, so that, for example, buddies who are at home can be easily separated from those who are in the office, and the user can send the last group an SMS with an invitation for a drink after office hours.
- *Personal context-aware diaries.* In a blog the focus shifts from providing context information in a timely manner to providing context overviews that are human-readable and informative. So, while in the buddy list the task is to show the last observed location of a buddy, in the blog the location time series needs to be summarized in a single sentence, say 'Today I travelled from Enschede (home) to Amsterdam and back.'. What is in the blog can be fully configured by the user and overnight the blog entries will be automatically generated based on the observed context information. However, one might have a blog with a daily paragraph describing travel during the day, the places visited, the local weather, the people encountered and the books read. Such a blog might be very interesting

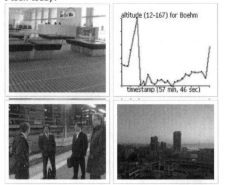

The cities that I visited today: Yokohama (4.5h), Tokyo (19.1h).
The buddies that I met today: Wagner (21.5h), Furusem (10.6h).

Figure 11-6.   A sample ContextWatcher blog

for relatives who live at a distance to browse, so that they can be brought up to date with your activities in a very unobtrusive manner. An example blog entry is shown in Figure 11-6.

Both methods of context sharing show how easy it is to keep in touch with friends, family and colleagues, using innovative mobile applications. The only requirement for the user is to have a ContextWatcher-enabled mobile phone with him, running 24 hours a day, 7 days a week. Buddy lists and blogs are updated automatically; the only effort for the user is to configure them according to his requirements.

11.2.2.2 Key ContextWatcher Features

The ContextWatcher application is modular in set-up, being able to combine different components and tabs from different developers, resulting in a dynamically configurable application set-up and automatic updates at component level. In this way we can offer multiple version of the same application (for example, light, target-group specific and full), the application is easy for new developers to extend and for existing developers to maintain. An impression of the architecture is given in Figure 11-7.

Interaction with the distributed network of context providers is via the mobile GPRS or UMTS network. Data read from the local sensors will be pushed to the remote context providers. These context providers can augment context information, reason with context information and share context information with authorized persons:

- *Augmenting*. This involves consulting other information services to provide additional information about a context parameter, for example to provide the product name

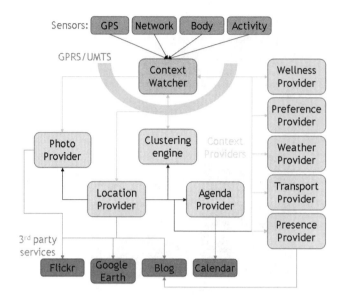

Figure 11-7. An impression of the architecture of the ContextWatcher application

and type for an EAN-13 bar code or the address information for a geo-location from a GPS.

- *Reasoning.* This involves providing situational information based on different context data streams, optionally from different people, for example to deduce that someone is currently in a business meeting from the fact that he is in the same room as other people, a few colleagues and one boss, with the data projector on.
- *Sharing.* This involves providing context information in raw or augmented form to other authorized persons, for example to buddies or to third-party services including blogs and photo albums.

For example, the interaction with the location provider is as follows: source data comes from either a GPS receiver (latitude, longitude) or from the network information service in the phone (cell id, network, country). This source data is pushed to the location provider at regular time intervals. The location provider augments the cell data with the geo-location using a large, operator-independent cell-id database such as http://client338.lab.telin.nl:8080/wasp/jsp/CellStats.jsp, and augments geo-location with address information using a geo-decoding service, such as http://www.mappoint.net, and stores the information in a database. This information is then used by other components, including the clustering engine, the photo provider and the weather provider.

The clustering engine analyses the location time series per person and tries to find the places frequently visited by that person, in order to be able to translate raw location information (geo-location or street) to concepts that are more meaningful to the user and his buddies. In conversations, we do not talk about absolute locations, rather about relative locations: my office, Marko's home or close to the church. These relative locations are easier to interpret and hence better to use when communicating about location in the buddy list or blog. The clustering engine analyses location streams every night, and it can extend existing clusters or find new ones, which are presented to the user by the ContextWatcher for naming. Next to a name the user can also relate a cluster to a location ontology that covers most concepts, including home, office, hotel, sport etc.

The photo provider stores pictures taken with the camera phone and automatically augments them with tags, titles and descriptions, based on the situation in which the photo was taken. The situation information is obtained from other context providers, which are queried for information around the time the picture was taken. In this way we can tag each picture with automatically recorded street and city information, the geo-position, the speed and direction of movement, the name of the location cluster (is this a home or an office picture), the local weather and the people who are nearby.

We have integrated the photo context provider with Flickr (http://www.flickr.com), one of the largest public image servers at the present time, where the context information is submitted as tags and the descriptive text is automatically generated, for example 'I was on a [business trip] together with [Henk] and [Bernd] in [Oulu] and I made this picture of the [Alexanderkatu] while travelling [on public transport] to the [summer school]', where all the information between brackets is auto-generated. This means that one action on the mobile phone is enough to send a richly described picture to a remote image server, enabling others to easily find pictures to their liking, for example by browsing the context tags to separate the home pictures from the office pictures. Rich-context queries become feasible, for example 'Provide me with all pictures together with Bernd in Oulu when it was snowing'.

The weather provider provides both current weather conditions and forecasts with a granularity at the city level. Since the city information can be obtained from the location provider, there is no longer any need for the user to specify the city. One click is enough to update the weather information for any current city. The weather information is then presented in the ContextWatcher and stored for later reference, so that for example the blog-generation engine can query the weather provider for the conditions along the location track of a certain user on a particular day.

Many features in addition to those described in this section are available, for example management of social relationships, sport scenarios with real-time body data, sound recording and sharing, bar-code recognition with the camera phone, usage logging etc. And the features are likely to be improved and extended in the near future.

11.2.3 McAnt

The existence of qualitative data is one prerequisite for ontology-based reasoning support. Often a mapping process from low- to high-level contextual data that is otherwise necessary can be made dispensable, by making use of already existing sources of high-level context information such as the data stored in personal information management (PIM) applications.

The McAnt prototype allows experimentation with context reasoning based on qualitative information from commonly available sources. It demonstrates how existing data can be made accessible to ontology reasoning and, as a side-product, how derived knowledge can be used to enhance applications. McAnt (Figure 11-8) has been developed as a Java application with a native Mac OS Cocoa interface and is linked to the back-end reasoning engine RACER,[12]

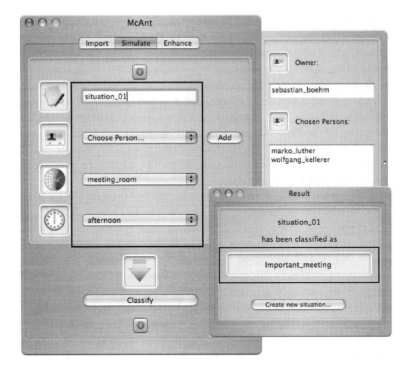

Figure 11-8. The McAnt simulation environment

the OS X address book and calendar applications, as well as the MobiLife OWL ontologies.

11.2.3.1 Retrieving Qualitative Information

Most people keep their electronic address book up to date and make use of calendar applications. Therefore, high-level context information is often readily available and can be accessed from standard PIM applications. If this were not possible, since qualitative information can only be processed in a sensible way using an ontology, the mapping of low-level data derived from sensor data to high-level context information would otherwise be a necessary preliminary step.

The information stored in an address book application often not only comprises a simple contact database but in addition offers the possibility of defining labels describing relationships between contact entries, including the owner (Figure 11-9). From the relationship definitions, individuals who are linked with object properties are created and classified in terms of appropriate T-box concepts such as 'Family' or 'Colleague'. Similarly, information managed using the calendar manager iCal is accessed and added to the ontology by McAnt. This time the high-level information gained consists of events, which can be associated with people attending a particular event and the location where it takes place. According to the concepts

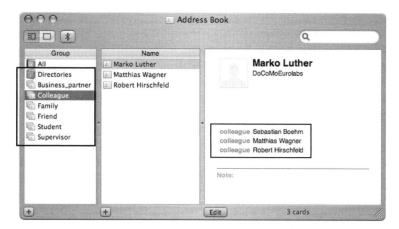

Figure 11-9. Mac OSX address book (www.apple.com)

defined in the schedule ontology, the reasoning system classifies these event instances with respect to their properties, based on the kind of event (Is it a business or private event?), its attendees as well as recurrence settings. So, if, for instance, the owner's supervisor attends a business meeting, it is categorized as an important business event.

11.2.3.2 Scenario Simulation

As well as being able to import high-level personal information into an existing T-Box structure, McAnt features the possibility of simulating certain scenario use cases. Through its simple user interface, one can easily create new scenarios by selecting, for example, persons attending, location and time using the available pull-down menus.

Rather than just organizing contextual information, ontology-based reasoning derives entailed knowledge. As a simple example, having defined transitive relationship properties enables automated role completion for individuals. This means that if two persons have been defined as colleagues of the owner, these two persons are also colleagues of one another.

The classification results are displayed in the results window according to the situation derived using the attached reasoning system. In this way, the use of ontologies for reasoning about the users' context can be demonstrated in a graspable manner, revealing the decisive entities that lead to the classification result.

11.2.3.3 Application Enhancement

The ongoing development in the field of consumer operating systems shows that intelligent search engines and consequently smarter ways of managing data are becoming more and more crucial. For instance, Apple has recently introduced the fourth major release of its desktop operating system Mac OS X Tiger, featuring Spotlight, a metadata-based search engine. This system-wide search technology has been consistently integrated in a variety of applications including Apple's standard PIM applications such as the Address Book. This means that the user can actually create so-called smart folders or, with respect to the Address Book, smart groups, which represent a search result in terms of certain criteria featuring simple conjunctions or disjunctions of attributes.

In this context, ontology-based reasoning mechanisms can further contribute to smarter ways of organizing different kinds of information. The information derived using logical reasoning based on an ontological background model can be used to enrich even those applications that originally provided the data. McAnt demonstrates this by enhancing the smart folders of Apple's address book application. These extended smart folders help to navigate through the address book database by classifying the contacts in terms of their relationship with the owner. For each defined sub-concept of the concept 'Person' in the ontology, a folder (such as 'Colleague') is created and populated by the individuals according to the classification result obtained from the reasoner based on the concept's specification (see Figure 11-8).

Unlike the rather limited possibilities for folder definitions in the unmodified Apple Address Book, our ontology-based approach provides the full expressiveness of OWL-DL. The latter can make use of the social relationships between the entities defined in the Address Book and therefore provide more sophisticated possibilities for creating smart folders. Furthermore, the logical consistency of the folder definitions is maintained, since conflicting definitions can be detected through the reasoner – something that is not possible otherwise.

11.2.4 MobiOWLS

MobiOWLS[13] is a new project in which we attempt to extend the OWL-S[14] upper ontology for services in order to describe services for mobile and ubiquitous computing. Our initial investigation concentrated on extending the OWL-S Profile to include crucial quality-of-service information and contextual information, which in our experience is required to locate the best service that satisfies the needs of the user. For example, we extended OWL-S with properties such as *Media*, which specifies the type of media, such as video or text, that is used to deliver the service, or the *CostModel* of the service, such as flat rate or a fee-per-use.

11.2.5 PERCI

The interaction between mobile devices and physical objects is receiving more and more attention since it is an intuitive way to request services from real-world objects. We currently see several solutions for the provision of such services, most of which are proprietary, designed for a special application area or interaction technique, and provide no generic concept for the description of services requested from real-world objects. On the other hand, ubiquitous environments with many tagged and networked objects – which are also often referred to as the *Internet of Things* – could provide generic standard methods for physical tagging and interfacing. In our project PERCI (Pervasive Service Interaction) we aim at leveraging Semantic Web technology to enrich and orchestrate such standard tagging and interaction methods. In particular, we study how Semantic Web service frameworks such as OWL-S could be extended towards ubiquitous service interaction.

11.3 Summary and Outlook

Mobile services will considerably enlarge the variety of applications accessible on the Web. We argue that an effective use of such services can only be achieved through extended service description, adequate personalization, advanced profiling and proactive service discovery and execution. In the future, base technologies such as federated online identity or XML encryption, which are currently in the process of standardization or are already on the market, will have to be combined with advanced profiling standards and description languages. Motivated by a concrete usage scenario, we have established a vision of a personal mobile Web based on existing technologies and emerging research efforts.

In our research – with the vision of contextual intelligence for mobile and ubiquitous service environments in mind – we are exploiting the usage of Semantic Web Technology within several practical projects, which we have briefly introduced above. These projects are either concerned with fundamental support for OWL-based development and/or with ontology-based services and applications in the mobile computing arena. We have presented our activities to date and have highlighted some of the challenges that we are facing right now. We have gained considerable experience in handling contextual information and the infrastructure that is required. Yet, despite the progress that has been made, we feel that there are still impressive challenges to tackle. Within the context of our projects we have also identified limitations and problematic issues in using Semantic Web Technology. In particular, in exploiting OWL we have found limitations in the language specification itself as well as in the tool support, which we reported elsewhere.[15]

12

From Personal Mobility to Mobile Personality

Matthias Wagner

12.1 Introduction

Difficulties in using and accessing new services have been the most frequently mentioned reasons for slow service acceptance in the past. For future mobile services to succeed, it is therefore critical that users are able to get intuitive and convenient access to the services they personally need in a given situation or context. In this chapter we introduce the concept of Mobile Personality, which builds on many concepts and technologies introduced earlier in this book. The idea is to allow the mobile user to develop her own online personality in terms of personal preference, usage and service profiles over time, as well as the offered services, to acquire a unique proactive behavior. This vision of adaptive personalized services is essentially based on advanced profiling and personalization concepts and context-aware computing, as well as flexible and evolvable service-support middleware. Through a practical use case, together with a detailed explanation and an interrelation of the essential enablers, we try to give an insight into the predicted transition from Personal Mobility to Mobile Personality.

It is assumed that systems beyond 3G (B3G) encompass heterogeneous access networks in order to provide the best available mobile connectivity to customers. These systems are not only considered to integrate several network platforms, but they also strongly encourage the vision of a substantial richness of services and applications. On the other hand – with the complexity of services constantly increasing – we can only grasp a vague impression of how most end-users will soon be confronted with a broad variety of services and multiple ways to combine them. Problems in accessing and using newly deployed services have already been the most frequently mentioned reasons for slow service adoption in the past. This may be even more the case for novel multimedia-type services and context-aware applications in

Towards 4G Technologies. Edited by Hendrik Berndt.
© 2008 John Wiley & Sons, Ltd.

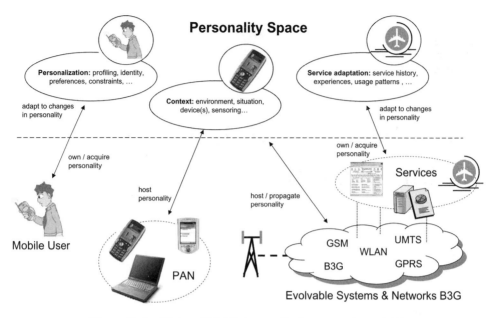

Figure 12-1. Vision of Mobile Personality in systems beyond 3G

the future. As a consequence, user- and context-sensitive provisioning of new services will be crucial.

This chapter (previously published in *Telenor Telektronikk Magazine*[1]) is motivated by our ongoing research activities on key concepts in enabling personalized service provisioning in systems beyond 3G. In projects such as the EURESCOM Project P1203 (http://www.eurescom. de) and IST MobiLife (http://www.ist-mobilife.org) we have identified, together with European operators, manufacturers and vendors, several system concepts and application domains that promise to be key drivers of future mobile communications systems. We suggest that effective use of future services can only be achieved through adequate personalization concepts and proactive service advertisement and adaptation. It will essentially be new concepts for service provisioning and deployment, together with new service paradigms, that attract customers and bring mobile communication beyond voice applications to the mass market. In the following we sketch out our concept of an evolving virtual personality, which is not limited to users, but can also be applied to other entities of a mobile system in order to implement adaptive personal services that develop their own initiative and spawn a Personality Space (Figure 12-1). As service deployment and usage develop over time, such a 'Mobile Personality' will allow different users to develop their personal preference profiles and typical service usage patterns according to their personal requirements, as well as allowing the services offered to acquire unique proactive behaviour.

12.2 Future User Profiling and Personalization

Personalized services and applications are considered to be among the most compelling features of mobile communication systems. They promise high customer benefit through the selection of specific services from a rapidly increasing diversity of mobile service offerings, and the adjustment of these services to the customer's individual needs. Taking the example of a business traveler, we have already sketched some of the key personalization issues in future mobile communication systems through application scenarios.[2,3] In the following we revisit these scenarios, particularly with regard to evolving a Mobile Personality.

Let us consider a sample user, named Michael, who plans and takes a business trip from Boston to Paris including many different steps and various responsibilities. Setting up all the necessary preparations is a complex matter and manually finding adequate services can already be quite time-consuming. However, communicating personal requirements and preferences to many services – which may not even be able to fulfil any of Michael's needs – will definitely become tedious. For the sake of simplicity, let us assume that, in Michael's case, the personalization and planning tasks for the business trip 'only' consist of:

- setting up the necessary transportation
- directions to the meeting place
- adaptation of services and user devices to changing environments

Obviously, Michael's trip will start with a flight from Boston to Paris. A flight booking agent, initiated, for example, through an interactive Web portal and later monitored on a mobile terminal, may try to buy a plane ticket according to Michael's personal preferences. In this case, all available airline services that allow the performance of the task of booking a flight need to be discovered. This could be done either manually through the travel portal or by using the activated travel agent. In either case, a flexible and intuitive way to model and express user preferences should be supported. In Michael's case, his user profile may contain hard constraints on the departure and arrival dates and on the class of the ticket, say Business Class, together with soft constraints on the preferred airline ('I prefer to fly with Air France rather than Delta'), the type of flight ('non-stop') and the air route ('as fast as possible'). Preferences associated with Michael are stored in his personal profile, which can either be stored on his primary personal device or be distributed over the network using a service such as 3GPP (Generic User Profile; http://www.3gpp.org).

In the same way that his air ticket is booked, Michael's travel agent may also take care of reserving a rental car at the arrival airport. Having successfully booked a flight, Michael finally arrives at the travel destination. He heads straight towards the airport's rental car centre, where he picks up his reserved car, following identification and authorization just via his mobile phone. His phone transparently and automatically initiates a discovery of available services. The in-car equipment synchronizes with the phone and automatically adjusts mirror, seat and heating to Michael's personal preferences. In addition to his preferred driving settings, Michael's phone discovers the car's built-in navigation system. Immediately the address for the meeting is transferred and the appropriate maps are loaded by the navigation system. Using a local traffic information service, the navigation system chooses a route and is able to predict the arrival time. Since the in-car system signals that there is enough time before the meeting starts, Michael decides to get some cash in the local currency. He accesses an ATM locator service, which shows the way to the ATMs closest to his current location that are able to charge his credit card at the lowest rate. Once Michael has selected an ATM, the navigation service has to adjust the route and collaborate with another service to find a nearby parking space.

Finally, Michael is guided to the meeting. Through his situation-aware communication environment, his preference of only being disturbed during meetings in the case of emergencies is activated. At the meeting room, the settings of Michael's communication devices are thus automatically adapted as the session starts. There is no longer any need to explicitly switch to another device profile. Even if there is no internal device profile available that meets the situation requirements, parameters from an external profile are transferred temporarily, carefully respecting emergency settings. During the meeting, Michael is able to transfer his video streaming session from his laptop to the built-in screen and conference system of the meeting room. In addition, other services such as a nearby printer and a video transcoder are discovered that support project work. All these devices are initialized with Michael's settings derived from his personal profile. Since an important partner in the meeting is called away by an emergency, the meeting has to be re-scheduled for later in the day taking the schedules of all meeting parties into account. Meanwhile, Michael may be interested to meet other people from the company. A scheduler service, available through the corporate WLAN, allows for short-term arrangements. It displays the availability of staff and administrative information such as room number or telephone, which are derived from the users' online profiles.

12.3 Enablers of a New Mobile Life Style

Various system and implementation aspects have to be considered in the engineering future service provisioning platforms that support the above scenario. The scenario already reveals a number of requirements that are essential for the realization of personalized service provisioning and a Mobile Personality. In particular, we consider the following requirements to be the most important:

- modelling the user through advanced profiling
- perception and modelling of the environment (context awareness)
- supporting a user-centred service discovery and service selection process based on the modelled information
- processing the modelled information and supporting its effect on the service execution, including service adaptation
- flexible service-support middleware that allows profile propagation and service transfer as well as personalization-driven service composition

In the following we will discuss the main building block for Mobile Personality.

12.3.1 Advanced Personalization Concepts

In our scenario, the 'personality' of a user is reflected in the set of personal profiles associated with him and it is passed to other users, network nodes and service providers. User modelling and profiling beyond device independence – covering user preferences and wishes – are essential for supporting such a virtual mobile personality. Recently profiling standards have been established for describing the service delivery context: the Composite Capabilities/ Preferences Profile (CC/PP) created by the W3C (http://www.w3.org/Mobile/CCPP), the User Agent Profile (UAProf) created by the Open Mobile Alliance (http://www. openmobilealliance.org – formerly the WAP Forum) as well as the Generic User Profile

(GUP) put forward by the 3GPP. They specify an XML- – and (in the case of CC/PP and UAProf) an RDF-based[4] framework – to address common needs for device-independent service access. Although they provide an interoperable basis for metadata descriptions of profile information, current profiling languages are not yet adequately suited for advanced profiling needs.

As shown in the scenario, semantic-based and cooperative service discovery and selection are integral parts of proactive services, that is, user needs and wishes will be identified as complex tasks, which are typically further divided into simpler sub-goals and are closely related to the user's context and environment. For user profiling languages there is much to learn from knowledge engineering and the database world, where the taxonomy or organization of profile elements is often referred to as a schema or ontology. We advocate that the design of future profiling languages for personalization can particularly benefit from the current approaches of the so-called Semantic Web,[5] where the layering of content descriptions has a similar quality: on top of XML, RDF provides a simple yet coherent structure for the expression of basic semantics and a foundation for different Web Ontology languages with a varying level of expressiveness[6] (see chapter 7). For services to be proactive, the sub-goals of a user's tasks have to be further explored and subsequently matched to adequate services. Matchmaking can be achieved with the help of advanced discovery and selection mechanisms that search the user's service environment according to his personal preferences.[7,8]

12.3.2 Context-aware Computing and Context Management

Context awareness is an attribute of a service that is capable of accessing, interpreting and manipulating knowledge of its environment and adapting the service behaviour accordingly.[9] To facilitate intelligent context-aware services that can develop their own personalities, we have developed a context management framework that allows the acquisition, management and processing of context information based on a flexible Web-Service-based framework.

This framework, developed within the MobiLife project and, in the following, referred to as the MobiLife context management framework (CMF),[10] enables the discovery and exchange of user context, as well as reasoning about context information. Its goal is to enable context information to flow easily from one provider to multiple context consumers, and from multiple providers to a single consumer, in order to build smart constellations of providers that finally can produce high-level situational information. This high-level information is then built up, based upon tiny bits of context information from heterogeneous sources, where heterogeneity has different dimensions, from syntax and semantics to transport, security, protocols and quality of context. The main tasks for the MobiLife context management framework are:

- enabling the discovery of context providers
- standardizing context exchange between providers and consumers
- supporting easy context reasoning by allowing reasoning components, such as a recommender, to be added to an application in a plug-and-play manner
- supporting the construction of different constellations of context providers to provide high-level situational information

MobiLife CMF is defined as a set of components, which are connected dynamically at run time, that together provide the relevant context information using sensing and interpretation mechanisms. Actual implementations of a context provider might range from simply wrapping a body sensor for a single user to a fully fledged inference reasoner that combines information gathered from other context providers for multiple users. CMF has been successfully used to implement and support different MobiLife applications including ContextWatcher.[11]

12.3.3 Flexible and Evolvable Service-support Middleware

Service-support middleware has to be capable of efficient session management including profile management and service mobility. The service session control as core part of the service-support middleware has to provide advanced service provisioning functions including those described above. We regard application-layer signalling to be acting as a coordination facility, supporting information exchange between the acting entities. To support advanced personalization concepts, we advocate a proxy-based service signalling approach.[12] Here, profile management, service session management and resource control are realized in separate interacting proxy servers that are on the end-to-end service signalling path. This concept is substantially enhanced to enable preference-based session management.[3] Preference-based session management makes use of profile information already in the proxy servers (for example network edge nodes) that are traversed by the service signalling messages and not only in the target application.

In addition, future systems have to be most flexible about user requirements and emergent services. To date, nearly all service architectures centre on the required functionality within a strictly layered system structure. Elsewhere, we have described an architecture that supports adaptation and evolution in systems B3G in general and on a middleware level in particular.[13] Such an adaptable environment allows personalization to be made effective for service execution by the flexible introduction and exchange of functional components, such as the context manager or the service selector. Programmable platforms on all system layers form the basis for the component management.

Adaptation and programmability of services should not be limited to content adaptation as it is in most of today's systems. Instead, it will involve modifications to behaviour (service logic)[14] and to service interaction and signalling as well. How dynamic service adaptation can benefit from dynamic programming environments such as Aspect-oriented Programming is further explained by Hirschfeld et al.[15,16]

12.4 Preference Patterns for Proactive Service Discovery

Using web-based services has already become an integral part of our everyday life. Semantic Web technology and the advent of universal and mobile access to Internet services will only add to the broad range of existing services on the web and provide additional features such as knowledge-based, location-aware or context-aware information. On the other hand, little work has been done so far to account explicitly for aspects of mobile computing in semantic service frameworks. Whereas much of the work in Semantic Web services discovery and composition has concentrated on the functionalities of the services, contextual information, personal preferences and more generally personalization are more pressing challenges in the mobile computing arena. In order to manage an increasing number of mobile services, it is essential that Semantic Web services standards explicitly support the needs of developers and users, such as the discovery and selection of services they personally need in a given situation or context.

In an early case study, we implemented MobiOnt and MobiXpl as a semantic toolbox to explore mobile user-cenered services on the Semantic Web.[5] Our vision is to take full advantage of future complex service offerings on limited client devices and to handle the need for personalized service discovery in mobile environments. MobiOnt and MobiXpl are early prototypes that have been realized and can be demonstrated as plug-ins to the Protégé knowledge workbench (http://protege.stanford.edu) as well as on Java-based mobile terminals that support tinySVG.[17] MobiXpl emulates different commercially available handsets, whereas MobiOnt encapsulates central preference-based matchmaking mechanisms. Implementations of MobiOnt as a central network component and MobiXpl as a Java-based client running on an actual phone are currently implemented.

12.4.1 A Concrete Usage Scenario

We study an extension of a previously published usage scenario[18] that addresses a future mobile Internet radio scenario. Internet radio has become increasingly popular in recent years with ever-growing numbers of Web radio stations and subscribers; see, for example, http://www.radio-locator.com/. In this context, personalized access to content is particularly important to accommodate both varying technical and personal user needs and preferences. For our use case we have modelled Internet radio stations as web services with varying service characteristics. Radio channels are described using an Internet radio ontology (a fragment of the ontology is shown in Figure 12-3) that consists of concepts that describe and classify web radio services in terms of program format, origin, audio format characteristics and a time-based classification of streamed audio content. This service ontology is then used for preference-based service discovery.

12.4.2 User Preferences

While the service ontology is browsed, service concepts with key relevance to the user can be selected and combined to form preferences. In our preference framework,[19] these (partially) ordered feature sets are directly handled without the use of any explicit quality or ranking values; user preferences are introduced as a special relation to the semantics of considering some object (or class) A as superior to another object (or class) B ('I like Music channels better than News stations'). Preferences indicate constraints that a service should fulfil to best meet its requirements. On the other hand, even if none of the indicated preferences are met, a match can still be possible. To manage multiple user preferences, complex preferences can be inductively constructed from a set of base preferences by means of preference constructors.[18,19]

Figure 12-2 shows an example of a combined preference from the radio scenario. Here a user has indicated that she generally prefers radio programs from Europe over those from Japan or America. Still, the latter two choices are her preferred choices over any other available program. As a result of the technical capabilities of her player, she also prefers MP3 encoding over Real. Further, she has specified that both base preferences are equally important to her.

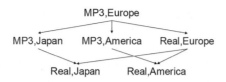

Figure 12-2. A user-defined preference ordering

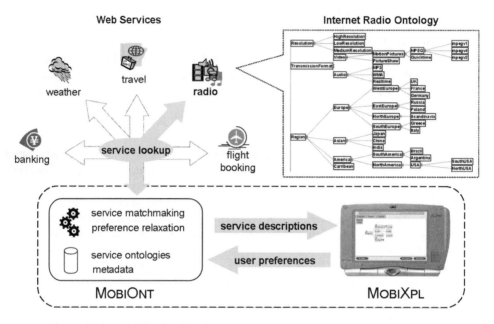

Figure 12-3. MobiOnt/MobiXpl – A testbed for mobile semantic-based services

12.4.3 Cooperative Service Discovery

User preferences constructed during preference building define a service request that ultimately needs to be mapped to the underlying service ontology. MobiOnt (Figure 12-3)
therefore implements a flexible discovery algorithm that can be extended through different
strategies. The goal of service discovery is to retrieve from the ontology those service
instances that represent the best matches to given preferences.

   The implemented preference-based service matching is performed along the lines of the
determined preference order to implement cooperative behaviour: if the search for a perfectly
matching radio station fails, the initial query is gradually relaxed along the path of the
(complex) preferences until a next-best match can be found. Thus, if during service discovery
in our example above no match could be found in European programs in MP3 encoding, the
next discovery step consists of trying to match radio stations that broadcast Japanese or
American programs in MP3 or European programs in Real. If neither of these two second-
best choices is available, any other program is matched. Further implementation and application aspects, as well as selective ontology browsing and preference building and mapping,
are further explored by Balke and Wagner.[8,20]

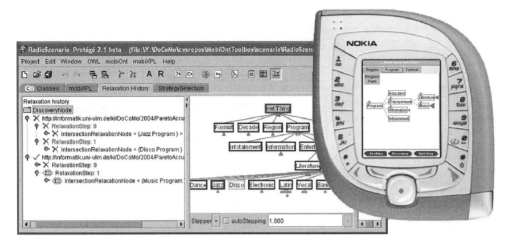

Figure 12-4. MobiXpl – exploring ontology-based service catalogues on the mobile device (Courtesy of NOKIA DSK)

12.4.4 MobiXpl – A User Interface for Preference-based Discovery

Parts of the Internet radio ontology are carefully exposed to the user through MobiXpl, the graphical front end to our framework (Figure 12-4). MobiXpl emulates different mobile terminals and consists of a mobile ontology browser with support for individual user views as well as an intuitive interface to user preferences. The idea is to display selected concepts and sub-ontologies depending only on the user's experience level and usage profile. While the service ontology is browsed, concepts that circumscribe services with key relevance to the user can be selected and combined to form user preferences. Subsequently, these preferences are used during the service discovery to implement cooperative behaviour: if the search for a perfectly matching radio station fails, the initial query is gradually relaxed along the lines of the determined preferences until a next-best match can be found. Both application aspects and selective ontology browsing, as well as preference building and mapping, are fully explored by Wagner *et al.*[18]

12.5 Towards a Mobile Personality

Expressing user preferences, wishes and dislikes in an intuitive way is crucial for the flexible provisioning of mobile services. With the vision of Mobile Personality in mind, we have developed essential concepts and solutions for the flexible matchmaking of evolving user preferences and services. Namely:

- an algorithm for preference-based negotiation and interaction with services[7]
- a mechanism for the use of semantically rich services descriptions, together with service usage patterns in aggregated service catalogues for cooperative service discovery[8,20]
- early prototypes that leverage preference-based matchmaking in semantically rich service catalogues for the mobile user.

In cases where the perfect match of the user's preferences to the available services is not possible in the given context, the next-nearest match has to be provided in a cooperative

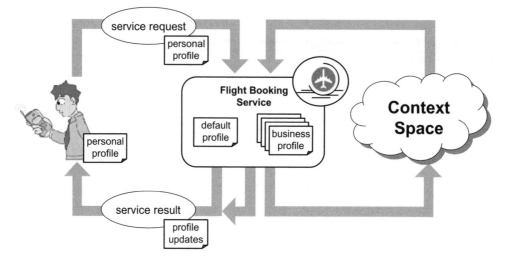

Figure 12-5. Cooperative service execution and context spaces

fashion. The outlined approach assumes that user preferences are modelled in terms of hard and soft constraints. Hard constraints model user preferences that definitely need to be matched during service discovery and selection, whereas soft constraints represent parts of a user profile or a user request that can be relaxed during the matchmaking to the available services. Our profiling concepts comprise a notion of usage patterns to express preferences of user groups and typical service invocation patterns. For example, a general preference from the travel domain could be that 'everyone prefers a short travelling time' (i.e. a departure date with maximum proximity to the arrival date is preferred). The iterative adaptation of these patterns to evolving user needs plays a key role in the concept of Mobile Personality. As indicated in Figure 12-5, user preferences are passed to the service provider for service matchmaking and execution. If no match between the personal profile and the default execution profile of the service is available, two basic conflict resolution strategies (and combinations thereof) are applicable: on the one hand the cooperative service execution may decide to relax user's preferences until a match can be made. On the other hand, the user may be associated with a typical group pattern, such as 'Business User', resulting in the enrichment of his request with additional preference data from the pattern.

Beyond the one-time usage, the life cycle of profile data is crucial in the Mobile Personality paradigm. For instance, when, for the first time, an intelligent agent reserves a rental car at Boston airport on behalf of our sample user Michael, it will do this based on some base preferences. However, on a revisit to Boston, Michael's user profile will already contain a usage history of services together with an experience report. Again, the left-hand side of Figure 12-5 illustrates this through the cycle of profile usage and profile update. Based on the user's profile and/or usage history – and maybe other usage histories, for example from Michael's colleagues or users with similar preferences and interest – a more informed decision can be made. In the case of Michael it is conceivable that the decision on a rental car agency, formerly price-based, will change based on experiences with the agency in the case of a car breakdown or based on unanticipated technical problems with adapting his mobile terminal to the in-car navigation system.

In Figure 12-5 we show a more detailed prospective of adaptive service interaction and the use of a context space for introspection and interaction by intelligent service agents: in the diagram, Michael initiates the flight booking service using a service request containing metadata representing his personal profile. The flight booking service, upon receiving the service request, processes the service request and promptly sends back the service result and the profile update. In addition to sending the service result to Michael, the flight booking service annotates the context space by writing the updated profile to the context space. Service agents with a similar interest in Michael, using the introspection capabilities of the context space, observe the annotation of the space and use logic to determine if the updates in the profile can be used for personalization. Through agent interaction, service agents, such as a group of rental car agents, reason that an additional profile update should be made available to Michael. The profile updates are annotated in the context space and the flight booking service is notified so that asynchronous delivery can be made to Michael.

12.6 Conclusion and Outlook

The evolution towards B3G systems and services bears the risk of increasingly confronting the customer with technology features instead of service aspects. We claim that similar problems in using and accessing new services have already led to slow service adoption in the past. As a consequence, users may not be able to fully understand novel services and to benefit from future applications. In this chapter – with customer acceptance and ease-of-use as the most important success factor for forthcoming telecommunication systems in mind – we have presented our vision and first steps towards supporting advanced personalization concepts under the umbrella of a 'Mobile Personality'.

The Mobile Personality paradigm is characterized through personal services, user preference and service profiles that evolve over time according to changes in the user's environment or context or in the service offerings themselves. The main building blocks of this personalization concept are advanced personalization concepts and context-aware computing, as well as highly flexible and evolvable system architectures. In our ongoing work emphasis has been put on generic approaches to context management, as well as on service discovery and selection that feature extendable semantic descriptions of services, users and devices.

13

Embracing the Real World through Services with Initiative

Hendrik Berndt

Throughout this book we have described methodologies, emerging solutions and promising insights that allow us to create a new and unprecedented way of Service Provision. By developing 4G mobile-environment-aware services we progress towards a new service universe. This new service universe is not only characterized by a great number of new applications, the beginnings of which can be observed with the emerging Web 2.0, but foremost by the enhancement of service semantics and meanings. Ontologies are efficient helpers for a common understanding of service behavior patterns, not only on their own but also in connection with others. Thus, composite services can be created at any desired level of sophistication and complexity. With this book all the building blocks are brought together that are needed for applications that adapt their behavior according to the complex environment in which they are used and that are deeply personalized in the best interest of the user. How the building blocks fit together is depicted in Figure 13-1.

Methods of advanced service discovery were outlined that allow for service selection suited to the user's real intentions and needs. Based on smart reasoning components, implicit information can be considered, and conflicts can be resolved in the process of service invocation by means of gradually alternating preferences and relaxing tight constraints on the user. Naturally, user guidance is desired if myriads of services are offered so that they are easily accessible for everyone. Usage patterns come in handy for mobile operators and service providers since they know best how services are used on a large scale and which features are appreciated most by the customers. Such knowledge allows them to create usage guidance mechanisms to ease the utilization of services. This is where our Services with Initiative are valued the most. A service provisioning environment which proactively takes the intentions and resources a user has at hand into consideration is helpful, and simplifies life considerably. The more a service understands the limitations of time, location, social settings, business

Towards 4G Technologies. Edited by Hendrik Berndt.
© 2008 John Wiley & Sons, Ltd.

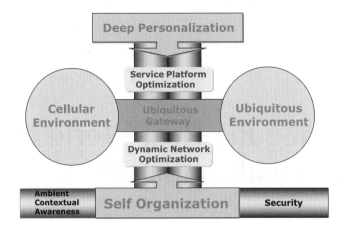

Figure 13-1. Building blocks for Services with Initiative

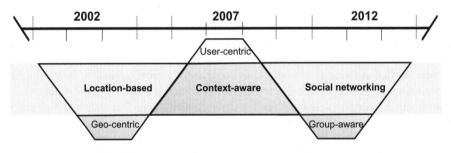

Figure 13-2. Service trends towards social networking

interests, available equipment etc., the better it can proactively configure itself to the current situation for the benefit of the customer.

In this book we showed how services are equipped so as to run in a ubiquitous networking environment, extending the reach to the real world, including all kinds of artifacts that have communication capabilities. More and more Near Field Communication services are used in the everyday lives of people within the Information Society, embracing the real world and shaping the Internet of Things. In general one can predict that as technology advances, users will get more and more empowered. Their role will change in many ways from business roles, such as from consumer to content and even service provider, from requirement settings regarding security and privacy to expectations of reliability and quality of service.

With services aware of the surroundings it becomes easy to find social components that match common interests, nearby friends and special events. This paves the way for an enormous set of Social Networked services. We observe the drift from location-based services towards mobile services, which are context-aware, and furthermore which become socially aware at the same time. All awareness levels are building upon each other. The trend towards social networking is depicted in Figure 13-2.

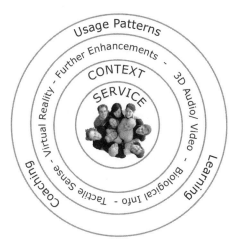

Figure 13-3. Vision of the Service Universe

Users who share a common view are finding each other, initiating peer groups, exchanging information, creating new services and organizing events. Self-presentations on everyday life logs on the one hand and community-based services that take on the challenges of the most burning societal needs such as environmental protection, healthcare etc. on the other hand are providing just a glimpse of social networking potential.

All the above-mentioned services and applications are part of our "service universe". We conclude this book with a picture (Figure 13-3) that outlines this service universe. It embraces the real world and provides meaningful services, embedded in and aware of the situations and environments that they are in. Enhancement of services with different kinds of virtual reality, biological information, tactile senses and other so far unused information sources is predicted for the not so distant future. An additional layer takes care of user guidance; coaching and learning based on usage patterns which includes support for efficient utilization of service functionalities based on resources available to the user.

We hope to see you soon in this Service Universe.

Appendix: References and Further Reading

Chapter 1: Introducing the 4G Mobile Adventure

References

[1] Inoue, Y., Lapierre, M. and Mossotto, C. (1999) *The TINA Book, A Cooperative Solution for a Competitive World*. Prentice Hall Europe.

[2] Prehofer, C., Kellerer, W., Hirschfeld, R., Berndt, H. and Kawamura, K. (2002) 'An Architecture Supporting Adaptation and Evolution in Fourth Generation Mobile Communication Systems' *Journal of Communications and Networks* 4(4).

[3] Niebert, Norbert (2007) *Ambient Networks: Co-operative Mobile Networking for the Wireless World*. John Wiley & Sons, Ltd.

Chapter 2: Mobile Communication Networks

References

[1] Manner, J. and Kojo, M. (eds) (2004) 'Mobility Related Terminology' RFC3753.

[2] Fenner, B., Handley, M., Holbrook, H. and Kouvelas, I. (2006) 'Protocol Independent Multicast – Sparse Mode (PIM-SM): Protocol Specification (Revised)' RFC4601.

[3] Adams, A., Nicholas, J. and Siadak, W. (2005) Protocol Independent Multicast – Dense Mode (PIM-DM): Protocol Specification (Revised)' RFC3973.

[4] 3rd Generation Partnership Project (2002) '3GPP TR 23.846 v6.1.0, Technical Specification Group Services and System Aspects, Multimedia/Multicast Service, Architecture and Functional Description, (Release 6)' December 2002.

[5] Kempf, J. and Austein, R. (eds) (2004) 'The Rise of the Middle and the Future of End-to-End: Reflections on the Evolution of the Internet Architecture' RFC3724.

[6] Clark, D.D., Wroclawski, J., Sollins, K. and Braden, R. (2002) 'Tussle in Cyberspace: Defining Tomorrow's Internet', *Proceedings of the ACM Sigcomm 2002*, Pittsburgh, PA.

[7] Perkins, C. (ed.) (2002) 'IP Mobility Support for IPv4' RFC3344.

[8] Johnson, D. *et al.* (2004) 'Mobility Support in IPv6' RFC3775.

[9] Campbell, A.T., Gomez, J., Kim, S., Wan, C.Y. and Turanyi, Z.R. (2002) 'Comparison of IP Micromobility Protocols' *IEEE Wireless Communications* 3(1), 72–82.

[10] Yumiba, H., Imai, K. and Yabusaki, M. (2001) 'IP-Based IMT Network Platform', *IEEE Personal Communications Magazine* 8(5), 18–23.

[11] Host Identity Protocol (hip), IETF Working Group, http://www.ietf.org/html.charters/hip-charter.html.

Towards 4G Technologies. Edited by Hendrik Berndt.
© 2008 John Wiley & Sons, Ltd.

[12] Braden, R., Berson, S., Herzog, S., Jamin, S. and Zhang, L. (1997) 'Resource ReSerVation Protocol (RSVP) – Version 1' RFC2205 (Standard), IETF.

[13] Davie, Bruce, Charny, Anna, Bennet, Jon, Benson, Kent, Le Boudec, Jean-Yves, Courtney, William, Davari, Shahram, Firoiu, Victor and Stiliadis, Dimitrios (2002) 'An Expedited Forwarding PHB.' RFC3246 (Standard), IETF.

[14] Heinanen, J., Baker, F., Weiss, W. and Wroclawski, J. (1999) 'Assured Forwarding PHB group' RFC2597 (Standard), IETF.

[15] NSIS, IETF Next Steps for Signalling (nsis) Working Group, http://www.ietf.org/html.charters/nsis-charter.html.

[16] Zhi-Li Zhang, Zhenhai Duan, Lixin Gao, Yiwei Thomas Hou (2000) 'Decoupling QoS Control from Core Routers: a Novel Bandwidth Broker Architecture for Scalable Support of Guaranteed Services' *ACM SIGCOMM Computer Communication Review, Proceedings of the Conference on Applications, Technologies, Architectures, and Protocols for Computer Communication*, **30**(4).

[17] Nichols, K., Jacobson, V. and Zhang, L. (1999) 'A Two-Bit Differentiated Services Architecture for the Internet' RFC2638 (Informational), IETF.

[18] Armitage, Grenville J. (2003) 'Revisiting IP QoS: Why Do We Care, What Have We Learned?' ACM SIGCOMM 2003 RIPQOS workshop report, *ACM SIGCOMM Computer Communication Review* **33**(5).

[19] Crowcroft, Jon, Hand, Steven, Mortier, Richard, Roscoe, Timothy and Warfield, Andrew (2003) 'QoS's Downfall: At the Bottom, or Not At All!' *Proceedings of the ACM SIGCOMM workshop on Revisiting IP QoS: What Have We Learned, Why Do We Care?* 25–27 August 2003, Karlsruhe, Germany.

[20] Manner, J. *et al.* (2002) 'Evaluation of Mobility and QoS Interaction,' *Computer Networks* **38**(February).

[21] Hillebrand, J., Prehofer, C., Bless, R. and Zitterbart, M. (2004) 'Quality-of-Service Signaling for Next-Generation IP-based Mobile Networks' *IEEE Communications Magazine*, June.

[22] Weiser, Mark (1991) 'The Computer for the Twenty-First Century' *Scientific American*. September 1991, pp. 94–104.

[23] Tschopp, D., Diggavi, S., Grossglauser, M. and Widmer, J. (2007) 'Robust Geo-routing based on Embeddings of Dynamic Wireless Networks, *Proceedings of IEEE Infocom, Anchorage, Alaska, USA* May 2007.

[24] Akyildiz, I.F., Wang, X. and Wang, W. (2005) 'Wireless Mesh Networks: A Survey' *Computer Networks Journal* **47**, 445–487.

[25] Gupta, P. and Kumar, P.R. (2000) 'The Capacity of Wireless Networks' *IEEE Transactions on Information Theory* **46**(2), 388–404.

[26] Bruno, R., Conti, M. and Gregori, E. (2005) 'Mesh Networks: Commodity Multihop ad hoc Networks' *IEEE Communications Magazine* **43**(3) 123–131.

[27] Niebert, N., Schieder, A., Abramowicz, H., Malmgren, G., Sachs, J., Horn, U., Prehofer, Ch. and Karl, H. (2004) 'Ambient Networks: An Architecture for Communication Networks beyond 3G' *IEEE Wireless Communications*, April 2004.

[28] Prehofer, C., Kellerer, W., Hirschfeld, R., Berndt, H. and Kawamura, K. (2002) 'An Architecture Supporting Adaptation and Evolution in Fourth Generation Mobile Communication Systems' *Journal of Communications and Networks* **4**(4).

[29] Decasper, D., Parulkar, G., Choi, S., DeHart, J., Wolf, T. and Plattner, B. (1999) 'A Scalable, High Performance Active Network Node', *IEEE Network*, January/February.

[30] Wetherall, David, Guttag, John and Tennenhouse, David (1999) 'ANTS: Network Services without the Red Tape' *IEEE Computer* **32**(4), 42–49.

Further Reading

Blake, S. Black, D., Carlson, M., Davies, E., Wang, Z., Weiss, W. (1998) 'An Architecture for Differentiated Services' RFC 2475, IETF.

Braden, R. (1994) 'Integrated Services in the Internet Architecture: an Overview' RFC 1633, IETF.

Eberspächer, J., Vögel, H.-J. and Bettstetter, C. (2001) *GSM – Switching, Services, and Protocols*, 2nd edn. Wiley.

Johansson, P., Kazantzidis, M., Kapoor, R. and Gerla, M. (2001) 'Bluetooth: An Enabler for Personal Area Networking' *IEEE Network*, September.

Karl, H. and Willig, A. (2005) *Protocols and Architectures for Wireless Sensor Networks*. Wiley.

Kellerer, W., Bettstetter, C., Schwingenschlögl, C., Sties, P., Steinberg, K.-E. and Vögel, H.-J. (2001) '(Auto)mobile Communication in a Heterogeneous and Converged World' *IEEE Personal Communications Magazine*, December.

Mann, S. (1997) 'Wearable Computing: A First Step toward Personal Imaging' *IEEE Computer*, February.
Mauve, M., Hartenstein, H., Füßler, H., Widmer, J. and Effelsberg, W. (2002) 'Positionsbasiertes Routing für die Kommunikation zwischen Fahrzeugen' *it + ti*, October.
Perkins, C.E. (ed.) (2001) *Ad Hoc Networking*, Addison Wesley.
Römer, K. and Mattern, F. (2004) 'The Design Space of Sensor Networks' *IEEE Wireless Communications*, December.
Vassis, D., Kormentzas, G., Rouskas, A. and Maglogiannis, I. (2005) The IEEE 802.11g Standard for High Data Rate WLANs' *IEEE Network*, May.
Vaughan-Nichols, S.J. (2004) 'Achieving Wireless Broadband with WiMax' *IEEE Computer*, June.
Walke, B.H. (2002) *Mobile Radio Networks. Networking, Protocols and Traffic Performance*, 2nd edn. Wiley.
Walke, B.H., Althoff, M.P. and Seidenberg, P. (2001) *UMTS – Ein Kurs*. Schlembach Fachverlag.
Weiser, M. (1991) 'The Computer for the Twenty-First Century' *Scientific American*, September.
Zheng, J. and Lee, M.J. (2004) 'Will IEEE 802.15.4 Make Ubiquitous Networking a Reality?: A Discussion on a Potential Low Power, Low Bit Rate Standard' *IEEE Communications Magazine*, June.
ZigBee Alliance (2005) *ZigBee Specification.*

Chapter 3: Mobile Service Systems

References

[1] ITU-T (1993) *Recommendation I.112: Vocabulary of Terms for ISDNs.*
[2] ITU-T (1993) *Recommendation Q.931: Digital Subscriber Signalling System No.1 (DSS1) – ISDN UNI Layer 3 Specification for Basic Call Control.*
[3] ITU-T (undated) *Recommendation Series Q.12xx: Intelligent Network.*
[4] Magedanz, T. and Popescu-Zeletin, R. (1996) *Intelligent Networks*. International Thomson Computer Press, London.
[5] OMG (2005) *Common Object Request Broker Architecture (CORBA/IIOP), v. 3.0.3.*
[6] Inoue, Y., Lapierre, M. and Mossotto, C. (1999) *The TINA Book – A Co-operative Solution for a Competitive World*. Prentice Hall Europe, London.
[7] Eberspächer, J., Vögel, H.-J. and Bettstetter, C. (2001) *GSM Switching, Services and Protocols*, 2nd edn. Wiley, 2001.
[8] Christensen, G., Florack, P. and Duncan, R. (2000) *Wireless Intelligent Networking*, Artech House.
[9] Holma, H. and Toskala A. (eds) (2001) *WCDMA for UMTS*. Wiley, Chichester.
[10] 3rd Generation Partnership Project (2002) *The Virtual Home Environment, version 5.1.0.*
[11] ITU-T (1998) *Recommendation H.323: Packet-Based Multimedia Communications System.*
[12] ITU-T (1998) *Recommendation H.450: Generic Functional Protocol for the Support of Supplementary Services in H.323.*
[13] Rosenberg, J., Schulzrinne, H., Camarillo, G., Johnston, A., Peterson, J., Sparks, R., Handley, M. and Schooler, E. (2002) 'SIP – Session Initiation Protocol' RFC3261, IETF, June 2002.
[14] Glasmann, J., Kellerer, W. and Müller, H. (2003) 'Service Architectures in H.323 and SIP – A Comparison' *IEEE Communication Surveys and Tutorials*, **5**(2).
[15] Lennox, J., Wu, X. and Schulzrinne, H. (2004) 'Call Processing Language (CPL): A Language for User Control of Internet Telephony Services' RFC3880, IETF.
[16] Handley, M. and Jacobson, V. (1998) 'SDP: Session Description Protocol' RFC2327, IETF.
[17] Camarillo, G. and Garcia-Martin, M. (2004) *The 3G IP Multimedia Subsystem*. Wiley.
[18] Cuervo, F., Greene, N., Rayhan, A., Huitema, C., Rosen, B. and Segers, J. (2000) 'Megaco Protocol Version 1.0' RFC 3015, IETF.
[19] The Parlay Group (2005) 'Parlay and OSA Technical Library' [http://www.parlay.org/].
[20] Open Mobile Alliance (2005) 'Wireless Application Protocol' [http://www.openmobilealliance.org/tech/affiliates/wap/wapindex.html].
[21] Natsuno, T. (2003) *I-Mode Strategy*. Wiley.
[22] Imai, K., Takita, W., Kano, S. and Kodate, A. (2005) 'An Extension of 4G Mobile Networks towards the Ubiquitous Real Space' *IEICE Transactions on Communications*, **E88-B**(7).
[23] Arbanowski, S., Ballon, P., David, K., Droegehorn, O., Eertink, H., Kellerer, W., van Kranenburg, H., Raatikainen, K. and Popescu-Zeletin, R. (2004) 'I-centric Communications: Personalization, Ambient Awareness, and Adaptability for Future Mobile Services' *IEEE Communications Magazine* **42**(9), 63–69.

[24] Schulzrinne, H. and Wedlund, E. (2000) 'Application-Layer Mobility using SIP' *ACM MC2R* **4**(3).

[25] Raman, B., Katz, R.H. and Joseph, A.D. (2000) 'Universal Inbox: Providing Extensible Personal Mobility and Service Mobility in an Integrated Communication Network' *Workshop on Mobile Computing Systems and Applications* (WMCSA'00).

[26] Shiaa, M.M. and Aagesen, F.A. (2002) 'Architectural Considerations for Personal Mobility in the Wireless Internet' *Personal Wireless Communication* (PWC 2002), Singapore.

[27] Kikuta, Y. *et al.* (2003) 'Design of Seamless Service Environment for Adaptive Service Transfer among Terminals' *8th International Workshop on Mobile Multimedia Communications* (MoMuC 2003), Munich.

[28] Song, H., Chu, H. and Kurakake, S. (2002) 'Browser Session Preservation and Migration' *11th International World Wide Web Conference* (WWW2002), Hawaii.

[29] Bettstetter, C., Kellerer, W. and Eberspächer, J. (2000) 'Personal Profile Mobility for Ubiquitous Service Usage, Version 1.0' *Book of Visions 2000*. ISTWireless Strategic Initiative (WSI).

[30] Hasekawa, M. (2003) 'Cross-Device Handover Using the Service Mobility Proxy' *Wireless Personal Multimedia Communications (WPMC03)*, Yokosuka, Japan.

[31] Shacham, R., Schulzrinne, H., Thakolsri, S. and Kellerer, W. (2005) 'The Virtual Device: Expanding Wireless Communication Services through Service Discovery and Session Mobility' *Proceedings of IEEE International Conference on Wireless and Mobile Computing, Networking and Communications*, WiMob'2005, Montreal, Canada, 22–24August.

[32] Shacham, R., Schulzrinne, H., Thakolsri, S. and Kellerer, W. (2007) 'Session Initiation Protocol (SIP) Session Mobility', IETF Internet Draft, November 2007, draft-shacham-sipping-session-mobility-05.txt – work in progress.

[33] Rosenberg, J., Peterson, J. Schulzrinne, H. and Camarillo G. (2004) 'Best Current Practices for Third Party Call Control (3pcc) in the Session Initiation Protocol (SIP)', RFC3725, IETF.

[34] Sparks, R. (2003) 'The Session Initiation Protocol (SIP) Refer Method', RFC 3515, IETF.

[35] Guttman, E., Perkins, C., Veizades, J. and Day, M. (1999) 'Service Location Protocol, Version 2' RFC2608, IETF.

Further Reading

Moerdijk, L. and Klostermann, A. (2003) 'Opening the Networks with Parlay/OSA: Standards and Aspects Behind the APIs' *IEEE Network Magazine* May/June.

Roach, A. (2002) 'Session Initiation Protocol (SIP) – Specific Event Notification' RFC3265, IETF.

Rosenberg, J. (2007) 'The Extensible Markup Language (XML) Configuration Access Protocol (XCAP)' RFC4825, IETF.

Shacham, R., Schulzrinne, H., Thakolsri, S. and Kellerer, W. (2004) 'An Architecture for Location-based Service Mobility Using the SIP Event Model' *Proceedings of ACM MobiSys 2004*, Boston.

Shacham, R., Schulzrinne, H., Thakolsri, S. and Kellerer, W. (2007) 'Ubiquitous Device Personalization: The Next Generation of IP Telephony' *ACM Transactions on Multimedia Computing, Communications, and Applications* **3**(2).

Steinmetz, R. and Wehrle, K. (2005) *Peer-to-Peer-Systems and Applications* LNCS 3485. Springer.

Wagner, M. and Kellerer, W. 'Web Services Selection for Distributed Composition of Multimedia Content' *Proceedings of ACM Multimedia 2004*, New York.

Chapter 4: Extension to Ubiquitous: Mobile Peer-To-Peer

References

[1] Sen, S. and Wang, J. (2004) 'Analyzing Peer-to-peer Traffic across Large Networks' *IEEE/ACM Transactions on Networking (TON)* **12**, 219–232.

[2] Iyer, S., Rowstron, A. and Druschel, P. (2002) 'SQUIRREL: A Decentralized, Peer-to-peer Web Cache. *Proceedings of Twenty-First ACM Symposium on Principles of Distributed Computing (PODC 2002)*, Monterey, CA.

[3] Parreira, J.X., Donato, D., Michel, S. and Weikum, G. (2006) 'Efficient and Decentralized PageRank Approximation in a Peer-to-Peer Web Search Network' *International Conference on Very Large Data Bases (VLDB)*, Seoul, Korea.

[4] Dabek, F., Kaashoek, M.F., Karger, D., Morris, R. and Stoica, I. (2001) 'Wide-area Cooperative Storage with CFS' *Proceedings of the 18th ACM Symposium on Operating Systems Principles (SOSP '01)*, Chateau Lake Louise, Banff, Canada.

[5] Stoica, I., Adkins, D., Zhuang, S., Shenker, S. and Surana, S. (2002) 'Internet Indirection Infrastructure' *Proceedings of the ACM SIGCOMM Conference*.

[6] Castro, M., Druschel, P., Kermarrec, A.-M., Nandi, A., Rowstron, A. and Singh, A. (2003) 'SplitStream: High-bandwidth Content Distribution in a Cooperative Environment. Second *International Workshop on Peer-to-Peer Systems* (IPTPS '03), Berkeley, CA.

[7] Kostic, D., Rodriguez, A., Albrecht, J. and Vahdat, A. (2003) 'Bullet: High Bandwidth Data Dissemination Using an Overlay Mesh' *Proceedings of the 19th ACM Symposium on Operating System Principles (SOSP 2003)*, New York.

[8] Caesar, M., Castro, M., Nightingale, E.B., O'Sheal, G. and Rowstron, A. (2006) 'Virtual Ring Routing: Network Routing Inspired by DHTs' In *Proceedings of the ACM SIGCOMM conference 2006*, Italy.

[9] Aberer, K., Alima, L.O., Ghodsi, A., Girdzijauskas, S., Hauswirth, M. and Haridi, S. (2005) 'The Essence of P2P: A Reference Architecture for Overlay Networks, P2P2005' *The 5th IEEE International Conference on Peer-to-Peer Computing*, Konstanz, Germany

[10] Despotovic, Z. (2005) *Building Trust-aware P2P Systems: From Trust and Reputation Management to Decentralized E-Commerce Applications*. PhD Thesis (no 3313), EPFL, Switzerland.

[11] Gnutella (2001) 'Clip2. The gnutella protocol specification v0.4 (document revision 1.2)' [http://www9.li mewire.com/developer/gnutella protocol 0.4.pdf].

[12] Lv, Q., Cao, P., Cohen, E., Li, K. and Shenker, S. (2002) 'Search and Replication in Unstructured Peer-to-peer Networks. *International Conference on Supercomputing*, New York.

[13] Stoica, I., Morris, R., Karger, D., Kaashoek, F. and Balakrishnan, H. (2001) 'Chord: A Scalable Peer-To-Peer Lookup Service for Internet Applications' *Proceedings of the 2001 ACM SIGCOMM Conference*, pp. 149–160.

[14] Ratnasamy, S., Francis, P., Handley, M., Karp, R. and Shenker, S. (2001) A Scalable Content-Addressable Network. *Proceedings of ACM SIGCOMM '01*, pp. 161–172.

[15] Rowstron, A. and Druschel, P. (2001) 'Pastry: Scalable, Distributed Object Location and Routing for Large-scale Peer-to-peer Systems' *IFIP/ACM International Conference on Distributed Systems Platforms (Middleware)*, pp. 329–350.

[16] Malkhi, D., Naor, M. and Ratajczak, D'. (2002) 'Viceroy: A Scalable and Dynamic Emulation of the Butterfly' *Proceedings of the 21nd ACM Symposium on Principles of Distributed Computing, PODC 2002*, pp. 183–192, Monterey, CA.

[17] Manku, G.S., Bawa, M. and Raghavan, P. (2003) 'Symphony: Distributed Hashing in a Small World' *Proceedings of the 4th USENIX Symposium on Internet Technologies and Systems (USITS 2003)*, Seattle, WA.

[18] Kleinberg, J. (2000) 'The Small-World Phenomenon: An Algorithmic Perspective' *Proceedings of the 32nd ACM Symposium on Theory of Computing (STOC 2000)*, pp. 163–170.

[19] Maymounkov, Petar and Mazières, David (2002) 'Kademlia: A Peer-to-peer Information System Based on the XOR Metric' *1st International Workshop on Peer-to-peer Systems (IPTPS2002)*, Boston, MA.

[20] Aberer, K. (2001) 'P-Grid: A Self-organizing Access Structure for P2P Information Systems' *Proceedings of the Sixth International Conference on Cooperative Information Systems (CoopIS 2001)*, Trento, Italy.

[21] Garcés-Erice, E.L., Biersack, W., Ross, K.W., Felber, P.A. and Urvoy-Keller, G. (2003) 'Hierarchical P2P Systems' *Proceedings of ACM/IFIP International Conference on Parallel and Distributed Computing (Euro-Par)*, pp. 1230–1239, Klagenfurt, Austria.

[22] Mizrak, A.T., Cheng, Y., Kumar, V. and Savage, S. (2003) 'Structured Superpeers: Leveraging Heterogeneity to Provide Constant-Time *Lookup*' *Proceedings of the Third IEEE Workshop on Internet Applications (WIAPP'03)*, pp. 104–111, San Jose, CA.

[23] Gummadi, P.K., Gummadi, R., Gribble, S.D., Ratnasamy, S., Shenker, S. and Stoica, I. (2003) 'The Impact of DHT Routing Geometry on Resilience and Proximity' In *Proceedings of the ACM SIGCOMM conference 2003*, Karlsruhe, Germany.

[24] Manku, G.S. (2003) 'Routing Networks for Distributed Hash Tables' *Proceedings of the 22nd ACM Symposium on Principles of Distributed Computing (PODC)*, pp. 133–142, Boston, MA.

[25] Steinmetz, R. and Wehrle, K. (2005) *Peer-to-Peer-Systems and Applications*, LNCS 3485. Springer.

[26] Kaashoek, M.F. and Karger, D.R. (2003) 'Koorde: A Simple Degree-optimal Distributed Hash Table' *Proceedings of the 2nd International Workshop on Peer-to-Peer Systems (IPTPS '03)*, Berkeley, CA.

[27] Rhea, S., Geels, D., Roscoe, T. and Kubiatowicz, J. (2004) 'Handling Churn in a DHT' *Proceedings of the 2004 USENIX Annual Technical Conference, (USENIX '04)*, Boston, MA.

[28] Harren, M., Hellerstein, J.M., Huebsch, R., Loo, B.T., Shenker, S. and Stoica, I. (2002) 'Complex Queries in DHT-based Peer-to-Peer Networks' *First International Workshop on Peer-to-Peer Systems (IPTPS '02)*, Cambridge, MA.

[29] Datta, A., Hauswirth, M., John, R., Schmidt, R. and Aberer, K. (2005) 'Range Queries in Trie-structured Overlays' *Proceedings of the 5th IEEE conference on P2P Computing*, Konstanz, Germany.

[30] Triantafillou, P. and Pitoura, T. (2003) 'Towards a Unifying Framework for Complex Query Processing over Structured Peer-to-Peer Data Networks' *VLDB '03 Workshop on Databases, Information Systems, and Peer-to-Peer Computing*, Germany.

[31] Ramabhadran, S., Ratnasamy, S., Hellerstein, J.M. and Shenker, S. (2004) 'Brief Announcement: Prefix Hash Tree' *Proceedings of ACM PODC*, St Johns, Canada.

[32] Klemm, F., Datta, A. and Aberer, K. (2004) 'A Query-Adaptive Partial Distributed Hash Table for Peer-to-Peer Systems' *International Workshop on Peer-to-Peer Computing & DataBases (P2P&DB 2004)*, Crete.

[33] Garcés-Erice, L., Felber, P., Biersack, E.W., Urvoy-Keller, G. and Ross, K.W. (2004) 'Data Indexing in Peer-to-Peer DHT Networks' *Proceedings of the 24th International Conference on Distributed Computing Systems (ICDCS 2004)*, pp. 200–208, Tokyo.

[34] Skobeltsyn, G., Hauswirth, M. and Aberer, K. (2005) 'Efficient Processing of XPath Queries with Structured Overlay Networks' *Proceedings of ODBASE'05*, Agia Napa, Cyprus.

[35] Lakshminarayanan, Karthik, Rao, Ananth, Stoica, Ion and Shenker, Scott (2005) 'End-host Controlled Multicast Routing', *Elsevier Computer Networks*, Special Issue on Overlay Distribution Structures and their Applications.

[36] Castro, M., Druschel, P., Kermarrec, A.-M. and Rowstron, A. (2003) 'Scalable Application-level Anycast for Highly Dynamic Groups' *NGC 2003*, Munich, Germany.

[37] Aekaterinidis, I. and Triantafillou, P. (2006) 'PastryStrings: A Comprehensive Content-Based Publish/Subscribe DHT Network' *26th IEEE International Conference on Distributed Computing and Systems (ICDCS 06)*, Portugal.

[38] Kellerer, W., Schollmeier, R. and Wehrle, K. (2005) 'Peer-to-peer in Mobile Environments. In Steinmetz, R. and Wehrle, K. (eds), LNCS Volume 3485. Springer.

[39] Kirk, P. (2003) 'The Gnutella Protocol Specification' [http://rfc-gnutella.sourceforge.net, v. 0.6].

[40] Zöls, S., Schollmeier, R., Kellerer, W. and Tarlano, A. (2005) 'The Hybrid Chord Protocol: A Peer-to-peer Lookup Service for Context-aware Mobile Applications' *Proceedings of 2005 International Conference on Networking (ICN'05)*, Réunion Island, France.

[41] Zöls, S., Schubert, S., Despotovic, Z. and Kellerer, W. (2006) 'Hybrid DHT Design for Mobile Environments' *5th International Workshop on Agents and P2P Computing (AP2PC 06)*, held at AAMAS 06, Japan.

[42] Zöls, S., Despotovic, Z. and Kellerer, W. (2005) 'Cost-based Analysis of Hierarchical DHT Design' *Proceedings of the 6th IEEE conference on P2P Computing*, Cambridge, UK.

[43] Gruber, I., Schollmeier, R. and Kellerer, W.' (2004) Performance Evaluation of the Mobile Peer-to-peer Protocol' *Proceedings of ACM/IEEE International Workshop on Global Peer-to-Peer Computing*, 19–22 April, Chicago, IL.

[44] Gruber, I., Schollmeier, R., Kellerer, W. (2006) 'Peer-to-peer Communication in Mobile Ad-Hoc Networks'. *Ad Hoc & Sensor Wireless Networks. An International Journal* 2(2.3).

[45] Johnson, D. and Maltz, D. (2001) 'Dynamic Source Routing in ad hoc Wireless Networks. In Perkins, C.E. (ed.), *Ad Hoc Networking*, pp. 139–172. Addison-Wesley.

[46] Resnick, P., Zeckhauser, R., Friedman, E. and Kuwabara, K. (2000) 'Reputation Systems' *Communications of the ACM*, **43**(12), 45–48.

[47] Kamvar, S.D., Schlosser, M.T. and Garcia-Molina, H. (2003) 'EigenRep: Reputation Management in P2P Networks' *Proceedings of the World Wide Web Conference*, Budapest.

[48] Despotovic, Z. and Aberer, K. (2004) 'A Probabilistic Approach to Predict Peers' Performance in P2P Networks' *Eighth International Workshop on Cooperative Information Agents*, Erfurt, Germany.

[49] Kreps, D. and Wilson, R. (1982) 'Reputation and Imperfect Information' *Journal of Economic Theory*, **27**, 253–279.

[50] Blanc, A., Liu, Y.-K. and Vahdat, A. (2005) 'Designing Incentives for Peer-to-peer Routing' *Proceedings of the IEEE Infocom Conference*, Miami, FL.

Chapter 5: Mobile Middleware

References

[1] Mattern, F. and Sturm, P. (2003) 'From Distributed Systems to Ubiquitous Computing' *Fachtagung Kommunikation in verteilten Systemen (KIVS)*, February, pp. 3–25.

[2] Bakken, D.E. (2003) 'Middleware' In Dasgupta, J. (ed.) *Encyclopaedia of Distributed Computing*, Kluwer.

[3] Gaddah, A. and Kunz, T. (2003) *A Survey of Middleware Paradigms for Mobile Computing*, Technical report sce-03-16, Carleton University, Ottawa.

[4] OMG (2004) 'CORBA Notification Service, Version 1.1' [http://www.omg.org].

[5] Sun Microsystems (2002) 'Java Message Service (JMS) version 1.1' [http://java.sun.com/products/jms].

[6] Cugola, G. and Di Nitto, E. (2001) 'Using a Publish/Subscribe Middleware to Support Mobile Computing' *Advanced Topic Workshop on Middleware for Mobile Computing*, Hiderberg, Germany.

[7] Smith, B. (1982) *Reflection and Semantics in a Procedural Programming Languages*. Report, MIT, Cambridge, MA.

[8] Schmidt, D.C. and Cleeland, C. (1999) 'Applying Patterns to Develop Extensible and Maintainable ORB Middleware' *IEEE Communication Magazine*, Special Issue on Design Patterns, April.

[9] Kon, F., Gill, B., Anand, M., Campbell, R. and Dennis Mickunas, M. (2000) 'Secure Dynamic Reconfiguration of Scalable CORBA Systems with Mobile Agents' *IEEE Joint Symposium on Agent Systems and Applications/Mobile Agents*, Zurich.

[10] Mascolo, C., Capra, L., Zachariadis, S. and Emmerich, W. (2002) 'XMIDDLE: A Data-Sharing Middleware for Mobile Computing' *Personal and Wireless Communications Journal* **21**(1).

[11] Capra, L., Mascolo, C., Zachariadis, S. and Emmerich, W. (2001) 'Towards a Mobile Computing Middleware: a Synergy of Reflection and Mobile Code Techniques' *8th IEEE Workshop on Future Trends of Distributed Computing Systems (FTDCS'2001)*, Bologna, Italy. IEEE Computer Society Press.

[12] Karjoth, G., Lange, D. and Oshima, M. (1997) 'A security model for Aglets' *IEEE Internet Computing Magazine*, July/August.

[13] Bellavista, P., Corradi, A. and Stefanelli, C. (2001) 'Mobile Agent Middleware for Mobile Computing' *IEEE Computer Magazine* **34**(3), 73–81.

[14] emporphia Ltd (undated) 'FIPA-OS' [http://www.emorphia.com/research/about.htm].

[15] Telecom Italia Lab. (undated) 'JADE' [http://jade.cselt.it].

[16] EU IST Project (undated) 'Lightweight Extensible Agent Platform (LEAP)' [http://leap.crm-paris.com].

[17] Project JXTA (undated) 'JXTA' [http://www.jxta.org].

[18] Steglich, S., Vaidya, R. N., Gimpeliovskaja, O., Arbanouski, S., Popesku-Zeletin, R., Sameshima, S., Kawano, K. (2003) 'I-Centric Services Based on Super Distributed Objects' *ISADS 2003 The Sixth International Symposium*.

[19] Apple Computer (2005) 'Bonjour' [http://www.apple.com/macosx/features/bonjour].

[20] IETF (2005) 'Dynamic Configuration of IPv4 Link-Local Addresses' RFC3927.

[21] Fenkam, P., Kinda, E., Dustar, S., Gall, H., Reif, G. (2002) 'Evaluation of a Publish/Subscribe System for Collaborative and Mobile Working' *The 11th International Workshop on Enabling Technologies, Infrastructure for Collaborative Enterprise* (WETICE'02), Pittsburgh, PA.

[22] Kon, F., Gill, B., Anand, M., Campbell, R. and Mickunas, M. Dennis (2000) 'Secure Dynamic Reconfiguration of Scalable CORBA Systems with Mobile Agents' *IEEE Joint Symposium on Agent Systems and Applications/Mobile Agents*, Zurich.

[23] The 3rd Generation Partnership Project (3GPP) (2005) [http://www.3gpp.org/.

[24] European Telecommunications Standards Institute Smart Card Platform (ETSI SCP) (undated) [http://portal.etsi.org/scp/summary.asp.

[25] EMV 2000 version 4.1 (undated) [http://www.emvco.com.

[26] Sun Microsystems (2003) 'Java Card version 2.2.1 [http://java.sun.com/products/javacard].

[27] MULTOS version 5 (2000) [http://www.multos.com.

[28] Octpus (2005) [http://www.octopuscards.com/enindex.jsp.

[29] Oyster (2005) [http://www.oystercard.com.

[30] RATP (2005) 'Navigo Pass' [http://www.ratp.fr/corpo/service/navigo.html].

[31] NTT DoCoMo (undated) 'i-mode FeliCa' [http://www.nttdocomo.com/corebiz/services/imode/felica.html]

[32] Sony (undated) 'FeliCa' [http://www.sony.net/Products/felica].
[33] Sun Microsystems (2005) 'Connected Limited Device Configuration' [http://java.sun.com/products/cldc].
[34] Finkenzeller, K. (1999) *RFID-Handbook*. Wiley.
[35] Nokia (undated) 'RFID in Brief' [http://www.nokia.com/nokia/0,,55738,00.html].
[36] Solarski, M., Strick, L., Motonaga, K., Noda, C. and Kellerer, W. (2004) 'Flexible Middleware Support for Future Mobile Services and Their Context-Aware Adaptation' *Proceedings of Lecture Notes in Computer Science*, pp. 281–292. Springer.
[37] Noda, C. and Walter, T. (2004) 'Smart Devices for Next Generation Mobile Services' *Proceedings of CASSIS*, Marseille, pp. 192–209. Springer.
[38] Blefari-Melazzi, N., Kellerer, W., Noda, C., Salsano, S. and Seidl, R. (2005) 'The Simplicity Project: Architecture Concept' IEEE/IPSJ SAINT2005. *The 2005 International Symposium on Applications and the Internet, workshop on 'Next Generation IP-based Service Platforms for Future Mobile Systems'*, Trento, Italy, 31 January–4 February.

Further Reading

Blair, G.S., Coulson, G., Andersen, A., Blair, L., Clarke, M., Costa, F., Duran-Limon, H., Fitzpatrick, T., Johnston, L., Moreira, R., Parlavantzas, N., Saikoski, K. (2001) 'The Design and Implementation of Open ORB v2' *IEEE DS Online*, Special Issue on Reflective Middleware.
Bologna University (undated) 'SOMA' [http://www-lia.deis.unibo.it/Research/SOMA].
International Organization for Standards (undated) [http://www.iso.org.
Padovitz, A., Loke, S.W., and Zaslavsky, A. (2003) 'Using the Publish/Subscribe Genre for Mobile Agents' *Proceedings of 1st German Conf. on Multiagent System Technology (MATES'03)*, Erfurt, Germany, September, pp. 180–192. Springer-Verlag.

Chapter 6: Cross-Layer Design – a New Paradigm for Optimization of Mobile Communication Systems

References

[1] Zimmermann, H. (1980) 'OSI Reference Model – The ISO Model of Architecture for Open Systems Interconnection' *IEEE Transactions on Communications* 28(4), 425–432.
[2] Kawadia, V. and Kumar, P. (2003) 'A Cautionary Perspective on Cross Layer Design' *IEEE Wireless Communication Magazine*, July.
[3] Shakkottai, S., Rappaport, T. and Karlsson, P. (2003) 'Cross-layer Design for Wireless Networks' *IEEE Communications Magazine*, October.
[4] Clark, D. and Tennenhouse, D. (1990) 'Architectural Considerations for a New Generation of Protocols *Computer Communication Review*, ACM SIGCOMM '90 Symposium.
[5] Haas, Z. (2001) 'Design Methodologies for Adaptive and Multimedia Networks' *IEEE Communication Magazine*, November.
[6] Rappaport, T., Annamalai, A., Buehrer, R. and Tranter, W. (2002) 'Wireless Communications: Past Events and a Future Perspective' *IEEE Communications Magazine* 40(5), 148–161.
[7] Adve, S., Harris, A., Hughes, C., Jones, D., Kravets, R., Nahrstedt, K., Sachs, D., Sasanka, R., Srinivasan, J. and Yuan, W. (2002) 'The Illinois GRACE Project: Global Resource Adaptation through Cooperation *Proceedings of the Workshop on Self-Healing, Adaptive, and Self-managed Systems* (held in conjunction with the 16th Annual ACM International Conference on Supercomputing), June.
[8] Ludwig, R. (1999) *A Case for Flow-adaptive Wireless Links* Technical Report UCB//CSD-99-1053. University of California at Berkeley.
[9] Sternad, M. (2002) *The Wireless IP Project*. RVK, Stockholm.
[10] Chen, K., Shah, H. and Nahrstedt, K. (2002) 'Cross-Layer Design for Data Accessibility in Mobile Ad-hoc Networks' *Wireless Personal Communications* 21(1).
[11] Yuen, W., Lee, H. and Andersen, T. (2002) 'A Simple and Effective Cross Layer Networking System for Mobile ad hoc Networks' *Proceedings of IEEE PIMRC*.

[12] Peng, Y., Khan, S., Steinbach, E., Sgroi, M. and Kellerer, W. (2005) 'Adaptive Resource Allocation and Frame Scheduling for Wireless Multi-User Video Streaming' *Proceedings of ICIP 2005.*

[13] Gross, J., Klaue, J., Karl, H. and Wolisz, A. (2004) 'Cross-Layer Optimization of OFDM Transmission Systems for MPEG-4 Video Streaming' *Computer Communications*, **27**, 1044–1055.

[14] Zhang, Q., Zhu, W. and Zhang, Y. (2002) A Cross-layer QoS-supporting Framework for Multimedia Delivery over Wireless Internet' *International Packet Video Workshop 2002.*

[15] Tupelly, R.S., Zhang, J. and Chong, E.K.P. (2003) 'Opportunistic Scheduling for Streaming Video in Wireless Networks' *Proceedings of the 37th Annual Conference on Information Sciences and Systems*, Baltimore, MD, 12–14 March.

[16] Xylomenos, G. and Polyzos, G. (1999) 'Link Layer support for Quality of Service on Wireless Internet Links' *IEEE Personal Communication Magazine* **6**(5), 52–60.

[17] Zhang, Y. and Cheng, L. (2003) 'Cross-Layer Optimization for Sensor Networks' *New York Metro Area Networking Workshop 2003*, New York City, 12 September.

[18] Khan, S., Sgroi, M., Peng, Y., Steinbach, E. and Kellerer, W. (2006) 'Application-driven Cross Layer Optimization for Video Streaming over Wireless Networks' *IEEE Communications Magazine*, Special Issue on Cross-Layer Protocol Engineering **44**(1).

[19] Krunz, M. and Tripathi, S.K. (1997) 'Exploiting the Temporal Structure of MPEG Video for the Reduction of Bandwidth Requirements' *Proceedings of INFOCOM 1997.*

[20] Tu, W., Kellerer, W. and Steinbach, E. (2004) 'Rate-Distortion Optimized Video Frame Dropping on Active Network Nodes' *Packet Video Workshop 2004*, Irvine, CA, 13–14 December.

[21] Khan, S., Sgroi, M., Steinbach, E. and Kellerer, W. (2005) 'Cross-Layer Optimization for Wireless Video Streaming – Performance and Cost' *Proceedings of ICME 2005.*

[22] Khan, S., Duhovnikov, S., Steinbach, E., Sgroi, M. and Kellerer, W. (2006) 'Application-driven Cross-layer Optimization for Mobile Multimedia Communication using a Common Application Layer Quality Metric' *Proceedings of 2nd IEEE International Symposium on Multimedia over Wireless (ISMW 2006)*, as part of International Wireless Communications and Mobile Computing Conference (IWCMC 2006), Vancouver 3–6 July 3–6.

Chapter 7: Ontologies

References

[1] Gruber, Th. (1993) 'A Translation Approach to Portable Ontology Specifications' *Knowledge Acquisition* **5**(2), 199–220.

[2] Staab, S. and Studer, R. (eds) (2004) *Handbook on Ontologies*. Springer.

[3] Baader, F., Calvanese, D., McGuinness, D., Nardi, D. and Patel-Schneider, P. (2003) *The Description Logic Handbook – Theory, Implementation and Applications*. Cambridge University Press.

[4] Schmidt-Schauß, M. and Smolka, G. 'Attributive Concept Descriptions with Complements' *Artificial Intelligence* **48**(1), 1–26.

[5] Sattler, U. (1996) 'A Concept Language Extended with Different Kinds of Transitive Roles' In Görz, G. and Hölldobler, S. (eds), *20. Deutsche Jahrestagung für Künstliche Intelligenz*, number 1137 in *Lecture Notes in Artificial Intelligence*. Springer.

[6] Baader, F. and Hanschke, Ph. (1991) 'A Schema for Integrating Concrete Domains into Concept Languages' *Proceedings of the 12th International Conference on Artificial Intelligence (IJCAI'91)*, pp. 452–457.

[7] Horrocks, I. (1997) *Optimising Tableaux Decision Procedures for Description Logics*. PhD thesis, University of Manchester.

[8] Horrocks, I. and Sattler, U. (1999) 'A Description Logic with Transitive and Inverse Roles and Role Hierarchies' *Journal of Logic and Computation* **9**(3), 385–410.

[9] Horrocks, I., Sattler, U. and Tobies, S. 'Practical Reasoning for Very Expressive Description Logics' *Logic Journal of the IGPL* **8**(3), 239–263.

[10] Horrocks, I. and Sattler, U. (2001) 'Ontology Reasoning in the SHOQ(D) Description Logic' In Nebel, B. (ed.), *Proceedings of the 17th International Joint Conference on Artificial Intelligence (IJCAI'01)*, pp. 199–204. Morgan Kaufmann.

[11] Horrocks, I., Li, L., Turi, D. and Bechhofer, S. (2004) The Instance Store: DL reasoning with large numbers of individuals. In Haarslev, V. and Möller, R. (eds), *International Workshop on Description Logics (DL'04)*, pp. 31–40. Whistler, British Columbia, Canada, June.

[12] Lutz, C., Areces, C., Horrocks, I. and Sattler, U. (2004) 'Keys, Nominals, and Concrete Domains *Journal of CEUR Workshop Proceedings* **49**, 170–179.

[13] Haarslev, V., Möller, R. and Wessel, M. (2005) 'Description Logic Inference Technology: Lessons Learned in the Trenches' In Horrocks, I. *et al.* (eds.), *Proceedings of the International Workshop on Description Logics (DL'05)*, pp. 160–167, July.

[14] Tobies, S. (2000) 'The Complexity of Reasoning with Cardinality Restrictions and Nominals in Expressive Description Logics' *Journal of Artificial Intelligence Research* **12**, 199–217.

[15] Horrocks, I. and Sattler, U. (2005) A Tableaux Decision Procedure for SHOIQ' *Proceedings of 19th International Joint Conference on Artificial Intelligence (IJCAI'05)*.

[16] Brachman, R. and Schmolze, J. (1985) 'An Overview of the KL-ONE Knowledge Representation System' *Cognitive Science* **9**(2), 171–216.

[17] MacGregor, R. (1991) 'Using a Description Classifier to Enhance Deductive Inference' *Proceedings of the 7th IEEE Conference on AI Applications*, pp. 141–147.

[18] Baader, F., Hollunder, B., Nebel, B., Profitlich, H.-J. and Franconi, E. (1994) 'An Empirical Analysis of Optimization Techniques for Terminological Representation Systems' *Applied Intelligence* **4**(2), 109–132

[19] Horrocks, I. (1998) 'Using an Expressive Description Logic: Fact or Fiction?' In Cohn, A.G., Schubert, L. and Shapiro, S.C. (eds), *Proceedings of the Sixth International Conference on Principles of Knowledge Representation and Reasoning (KR'98)*, pp. 636–647, San Francisco, CA, June. Morgan Kaufmann.

[20] Haarslev, V. and Möller, R. (2001) 'Racer System Description' *International Joint Conference on Automated Reasoning (IJCAR'01)* 18–23 June, Siena, Italy. Springer.

[21] Sirin, E. and Parsia, B. (2004) 'Pellet: An OWL DL Reasoner' In Haarslev, V. and Möller, R. (eds), *International Workshop on Description Logics (DL'04)*, pp. 212–213. Whistler, British Columbia, Canada, June.

[22] Tsarkov, D. and Horrocks, I. (2004) 'Efficient Reasoning with Range and Domain Constraints. In Haarslev, V. and Möller, R. (eds), *International Workshop on Description Logics (DL'04)*, pp. 41–50. Whistler, British Columbia, Canada, June.

[23] Horrocks, I. (2005) 'Applications of Description Logics: State of the Art and Research Challenges' *Proceeedings of the 13th Inerntional Conference on Conceptual Structures (ICCS'05)*.

[24] Baader, F., Horrocks, I. and Sattler, U. (2003) 'Description Logics as Ontology Language for the Semantic Web' In Hutter, D. and Stephan, W. (eds), *Festschrift in Honor of Jörg H. Siekmann*. Springer.

[25] Berners-Lee, T., Hendler, J. and Lassila, O. (2001) 'The Semantic Web' *Scientific American* **284**(5), 34–43.

[26] Bechhofer, S., van Harmelen, F., Hendler, J., Horrocks, I., McGuinness, D., Patel-Schneider, P.F. and Andrea Stein, L. (2004) *OWL Web Ontology Language Reference* W3C Recommendation. The Worldwide Web Consortium.

[27] Horrocks, I. (2002) 'DAML+OIL: a Description Logic for the Semantic Web' *Bulletin of the IEEE Computer Society Technical Committee on Data Engineering* **25**(1), 4–9.

[28] Horrocks, I., Patel-Schneider, P. and van Harmelen, F. (2003) 'From SHIQ and RDF to OWL: The Making of a Web Ontology Language' *Journal of Web Semantics*, **1**(1).

[29] Patel-Schneider, P.F., Hayes, P. and Horrocks, I. (2004) *OWL Web Ontology Language Semantics and Abstract Syntax* W3C recommendation. The Worldwide Web Consortium.

[30] McGuinness, D. and van Harmelen, F. (2004) OWL *Web Ontology Language Overview* W3C Recommendation. The Worldwide Web Consortium.

[31] Tessaris, S. (2001) 'Querying Expressive DLs' *Proceedings of the 2001 International Description Logics Workshop (DL'01)*.

[32] Horrocks, I. and Tessaris, S. (2002) 'Querying the Semantic Web: a Formal Approach. In Horrocks, Ian and Hendler, James (eds), *Proceedings of the 2002 International Semantic Web Conference (ISWC'02)*, Volume 2342 of *Lecture Notes in Computer Science*, pp. 177–191. Springer.

[33] Fikes, R., Hayes, P. and Horrocks, I. (2004) 'OWL-QL – A Language for Deductive Query Answering on the Semantic Web *Journal of Web Semantics* **2**(1), 19–29.

[34] Wessel, M. and Möller, R. (2005) A high performance semantic web query answering engine. In Horrocks, I., Sattler, U. and Wolter, F. (eds) *International Workshop on Description Logics (DL'05)*, Edinburgh, Scotland, July, pp. 584–595. National e-Science Centre.

[35] Karvounarakis, G., Alexaki, S., Christophides, V., Plexousakis, D. and Scholl, M. (2003) 'Querying the Semantic Web with RQL' *Computer Networks* **42**(5), 617–640.

[36] Seaborne, A. (2004) *RDQL – a Query Language for RDF* W3C member submission. The Worldwide Web Consortium.

[37] Sintek, M. and Decker, St. (2002) 'TRIPLE – a Query, Inference, and Transformation Language for the Semantic Web' *International Semantic Web Conference (ISWC)*, Sardinia, June.

[38] Haase, P., Broekstra, J., Eberhart, A. and Volz, R. (2004) 'A Comparison of RDF Query Languages *Proceedings of the Third International Semantic Web Conference (ISWC'04)*, Hiroshima, Japan.

[39] Preece, A. (1996) 'Validating Dynamic Properties of Rule-Based Systems' *Journal of Human-Computer Studies* **44**, 145–169.

[40] Winston, P. (1994) *Artificial Intelligence*. Addison-Wesley.

[41] McBride, B. (2001) 'Jena: Implementing the RDF Model and Syntax Specification' *Proceedings of the 2nd International Workshop on the Semantic Web (SemWeb'01)*, Hongkong, May.

[42] Bruijn, J. de and Fensel, D. (2005) *OWL-. WSML Deliverable* WSML Working Draft D20.1 v0.2, 6 January. Digital Enterprise Research Institute (DERI).

[43] Schmidt-Schauß, M. (1989) 'Subsumption in KL-ONE is Undecidable' *Proceedings of the 1st Int. Conf. on the Principles of Knowledge Representation and Reasoning (KR'89)*, pp. 421–431. Morgan Kaufmann.

[44] Motik, B., Sattler, U. and Struder, R. (2004) 'Query Answering for OWL DL with Rules' *Proceedings of the Third International Semantic Web Conference (ISWC'04)*, Hiroshima, Japan, 2004. Defines the decidable DL-safe fragment of the Semantic Web Rule Language (SWRL).

[45] Grosof, B.N., Horrocks, I., Volz, R. and Decker, St. (2003) 'Description Logic Programs: Combining logic programs with Description Logic' *Procedings of the Twelfth International World Wide Web Conference (WWW 2003)*, pages 48–57. ACM, 2003.

[46] Levy, A. and Rousset, M.-Ch. (1998) 'Combining Horn Rules and Description Logics in CARIN' *Artificial Intelligence* **104**(1–2), 165–209.

[47] Horrocks, I. and Patel-Schneider, P. (2004) 'A Proposal for an OWL Rules Language' *Proceedings of the Thirteenth International World Wide Web Conference (WWW'04)*, pp. 723–731. ACM.

[48] Grau, B., Parsia, B. and Sirin, E. (2004) 'Working with Multiple Ontologies on the Semantic Web' *Proceedings of the Third International Semantic Web Conference (ISWC'04)*, Hiroshima, Japan. Springer.

[49] Kalyanpur, A., Parsia, B. and Hendler, J. (2005) 'A Tool for Working with Web Ontologies' *International Journal on Semantic Web and Information Systems* **1**(1), 36–49.

[50] Heflin, J. (2004) *OWL Web Ontology Language Use Cases and Requirements* W3C Recommendation. The Worldwide Web Consortium [http://www.w3.org/TR/webont-req].

[51] Heflin, J. and Muñoz-Avila, H. (2004) *Integrating HTN Planning and Semantic Web Ontologies for Efficient Information Integration* Technical Report LU-CSE-04-002. Department of Computer Science and Engineering, Lehigh University.

[52] Liebig, Th., Pfeifer, H. and von Henke, F. (2004) 'Reasoning Services for an OWL Authoring Tool: An Experience Report' In Haarslev, V. and Möller, R. (eds), *International Workshop on Description Logics (DL2004)*, pp. 79–82, Whistler, British Columbia, Canada, June.

[53] Bechhofer, S. (2003) 'The DIG Description Logic Interface: DIG/1.1' *Proceedings of the 2003 International Workshop on Description Logics (DL'03)*, Rome, Italy, June.

[54] Chen, C., Haarslev, V. and Wang, J. (2005) 'LAS: Extending Racer by a large abox store' In Horrocks, I., Sattler, U. and Wolter, F. (eds) *International Workshop on Description Logics (DL'05)*, Edinburgh, Scotland, July, pp. 200–207. National e-Science Centre.

[55] Gruber, Th. (1995) 'Towards Principles for the Design of Ontologies used for Knowledge Sharing' *International Journal of Human-Computer Studies* **43**, 907–928.

[56] Zhdanova, A.V. and Keller, U. (2005) 'Choosing an Ontology Language' *Proceedings of the Second World Enformatika Congress (WEC'05)*, pp. 47–50, Istanbul, Turkey, February.

[57] Martin, Ph. (2000) 'Conventions and Notations for Knowledge Representation and Retrieval' In B. Ganter and G.W. Mineau (eds), *Proceedings of the 8th International Conference on Conceptual Structures (ICCS'00)*, LNAI, vol. 1867, pp. 41–54, August. Springer-Verlag.

[58] Guha, R.V. and Bray, T. (1997) *Meta Content Framework using XML* W3C Technical Report. World Wide Web Consortium.

[59] Masolo, C., Borgo, S., Gangemi, A., Guarino, N. and Oltramari, A. (2003) *Ontology Library* WonderWeb Deliverable D18, EU-Project IST-2001-33052.

[60] Akkermans, H., Brown, M., Bouladoux, J.-M., Dieng, R., Ding, Y., Gómez-Pérez, A., *et al.* (2002) *Successful Scenarios for Ontology-based Applications* EU-Project IST-2000-29243 OntoWeb, Deliverable D21.

[61] Noy, N.F., Sintek, M., Decker, S., Crubezy, M., Fergerson, R.W. and Musen, M.A. (2001) 'Creating Semantic Web contents with Protégé-2000' *IEEE Intelligent Systems* **16**(2), 60–71.

[62] Gangemi, A., Guarino, N., Masolo, C., Oltramari, A. and Schneider, L. (2002) 'Sweetening Ontologies with DOLCE' In A. Gomez-Perez and V.R. Benjaminis (eds), *Proceedings of the 13th International Conference on Knowledge Engineering and Knowledge Management. Ontologies and the Semantic Web*, Vol. 2473, pp. 166–181. Springer.

[63] Bateman, J. and Farrar, S. (2004) 'Towards a Generic Foundation for Spatial Ontology' In *Proceedings of the 4th International Conference on Formal Ontology in Information Systems (FOIS'04)*, Torino, Italy.

[64] Tonti, G., Bradshaw, J.M., Jeffers, R., Montanari, R., Suri, N., Uszok, A. (2003) 'Semantic Web Languages for Policy Representation and Reasoning: A Comparison of KAoS, Rei, and Ponder' In Fensel, D., Sycara, K.P. and Mylopoulos, J. (eds), *Proceedings of the 2nd International Semantic Web Conference (ISWC'03)*, LNCS, October. Springer.

[65] Ding, L., Zhou, L., Finin, T. and Joshi, A. (2005) 'How the Semantic Web is being Used: An Analysis of FOAF' *Proceedings of the 38th International Conference on System Sciences*, January.

[66] Chen, H., Perich, F., Finin, T. and Joshi, A. (2004) 'SOUPA: Standard Ontology for Ubiquitous and Pervasive Applications' *International Conference on Mobile and Ubiquitous Systems: Networking and Services*, pp. 258–267, Boston, MA, August.

[67] Hobbs, J.R. and Pan, F. (2004) 'An Ontology of Time for the Semantic Web' *ACM Transactions on Asian Language Information Processing* **3**(1), 66–85.

[68] Niles, I. and Pease, A. (2001) 'Towards a Standard Upper Ontology' In Welty, Chris and Smith, Barry (eds), *Proceedings of the 2nd International Conference on Formal Ontology in Information Systems (FOIS'01)*, Ogunquit, ME.

[69] Chen, H., Finin, T. and Joshi, A. (2004) 'An Ontology for Context-aware Pervasive Computing Environments' *Knowledge Engineering Review*, Special Issue on Ontologies for Distributed Systems **18**(3), 197–207.

[70] Wang, X., Zhang, D., Dong, J., Chin, Ch. and Hettiarachchi, S. (2004) 'Semantic Space: A Semantic Web Infrastructure for Smart Spaces' *IEEE Pervasive Computing*, **3**(3), 32–39.

[71] Martin, D., Burstein, M., Hobbs, J., Lassila, O., McDermott, D., McIlraith, S., Narayanan, S., Paolucci, M., Parsia, B., Payne, T., Sirin, E., Srinivasan, N. and Sycara, K. (2004) *OWL-S: Semantic Markup for Web Services* W3C Member Submission.

[72] Semy, S., Pulvermacher, M. and Obrst, L. (2004) *Towards the Use of an Upper Ontology for U.S. Government and Military Domains: An Evaluation.* Technical Report TR-04-0603, MITRE, Bedford, MA.

[73] Farrar, S. and Bateman, J. (2004) *General Ontology Baseline.* OntoSpace German Project on Spatial Cognition, SFB/TR8. Deliverable D1, University of Bremen.

[74] Horrocks, I. and Voronkov, A. (2006) 'Reasoning Support for Expressive Ontology Languages Using a Theorem Prover' *Proceedings of the Fourth International Symposium on Foundations of Information and Knowledge Systems (FoIKS)*, LNCS 3861, pp. 201–218, Springer.

[75] Aftelak, A., Häyrynen, A., Klemettinen, M. and Steglich, S. (2004) 'MobiLife: Applications and Services for the User-centric Wireless World' *IST Mobile and Wireless Communications Summit*, Lyon, France, June.

[76] Floréen, P., Przybilski, M., Nurmi, P., Koolwaaij, J., Tarlano, A., Wagner, M., Luther, M., Bataille, F., Boussard, M., Mrohs, B. and Lau, S. (2005) 'Towards a Context Management Framework for MobiLife' *Proceedings of the 14th IST Mobile and Wireless Communication Summit (MOWICON'05)*, Dresden, Germany, 29–23 June 2005.

[77] Dawson, F. and Howes, T. (1998) *vCard.* The Internet Society.

[78] Brickley, D. and Miller, Li. (2005) *FOAF Vocabulary Specification.* Namespace Document.

[79] Mika, P. and Gangemi, A. (2004) 'Descriptions of Social Relations' *Proceedings of the 1st Workshop on Friend of a Friend, Social Networking and the Semantic Web*, Galway, Ireland, September.

[80] Pan, F. and Hobbs, J. (2004) 'Time in OWL-S' *Proceedings of the AAAI Symposium.*

[81] Allen, J.F. (1983) 'Maintaining Knowledge about Temporal Intervals' *Communications of the ACM*, **26**(11), 832–843.

[82] Randell, D.A., Cui, Z. and Cohn, Anthony G. (1992) 'A Spatial Logic based on Regions and Connections' *Proceedings of the Third International (KR'92)*, pp. 165–176. Morgan Kaufman.

[83] Liebig, Th. and Noppens, O. (2004) 'OntoTrack: Combining Browsing and Editing with Reasoning and Explaining for OWL Lite Ontologies' *Proceedings of the 3rd International Semantic Web Conference (ISWC'04)*, Hiroshima, Japan, November.

[84] Knublauch, H., Fergerson, R.W., Noy, N.F. and Musen, M.A. (2004) 'The Protégé OWL Plugin: An Open Development Environment for Semantic Web Applications' *Proceedings of the Third International Semantic Web Conference (ISWC'04)*, Hiroshima, Japan.

[85] Grau, B., Parsia, B., Sirin, E. and Kalyanpur, A. (2005) 'Automatic Partitioning of OWL Ontologies using *E*-connections' In Horrocks, I., Sattler, U. and Wolter, F. (eds) *International Workshop on Description Logics (DL'05)*, Edinburgh, Scotland, July. National e-Science Centre.

[86] Grau, B., Parsia, B. and Sirin, E. (2006) 'Combining OWL Ontologies using *E*-connections' *Journal of Web Semantics* **4**(1), 40–59.

[87] Kalyanpur, A., Parsia, B. and Sirin, E. Black box techniques for debugging unsatisfiable concepts' In Horrocks, I., Sattler, U. and Wolter, F. (eds) *International Workshop on Description Logics (DL'05)*, Edinburgh, Scotland, July. National e-Science Centre.

[88] Parsia, B., Sirin, E. and Kalyanpur, A. (2005) 'Debugging OWL Ontologies' *Proceedings of the 14th International World Wide Web Conference (WWW'05)*, May.

[89] Liebig, Th. and Halfmann, M. (2005) 'Explaining Subsumption in $\mathcal{ALEHFR}$+ Tboxes. In Horrocks, I., Sattler, U. and Wolter, F. (eds) *International Workshop on Description Logics (DL'05)*, Edinburgh, Scotland, July. National e-Science Centre.

Further Reading

Luther, M., Mrohs, B., Wagner, M., Steglich, S. and Kellerer, W. (2005) 'Situational Reasoning – a Practical OWL Use Case' *Proceedings of the 7th International Symposium on Autonomous Decentralized Systems (ISADS'05)*, Chengdu, China, April.

Mrohs, B., Luther, M. and Vaidya, R. (2005) 'Context-aware Presence Management' *Proceedings of the Workshop on Context Awareness for Proactive Systems (CAPS'05)*, pp. 100–103, Helsinki, Finland, June.

Mrohs, B., Luther, M., Vaidya, R., Wagner, M., Steglich, S., Kellerer, W. and Arbanowski, S. (2005) 'OWL-SF – a Distributed Semantic Service Framework' *Proceedings of the Workshop on Context Awareness for Proactive Systems (CAPS'05)*, pp. 67–77, Helsinki, Finland, June.

Chapter 8: Semantic Services

References

[1] NTT DoCoMo Inc. (2002) *I-Mode Service Guideline*, Version 1.2.0, 4 March [http://www.nttdocomo.com/technologies/present/imodetechnology/disclaimer.html].

[2] Booth, D., Liu, C. (2007) Web Services Description Language (WSDL) Version 2.0 Part 0: Primer; W3C Recommendation 26 June 2007 [http://www.w3.org/TR/wsdl20-primer]

[3] Mitra, N., Lafon, Y. (2007) SOAP Version 1.2 Part 0: Primer (Second Edition) [http://www.w3.org/TR/soap12-part0/]

[4] OASIS (2000) *The UDDI Technical White Paper* Technical report. OASIS.

[5] Booth, D., Haas, H., McCabe, F., Newcomer, E., Champion, M., Ferris, C. and Orchard, D. (2004) *Web Services Architecture*. W3C Working Group Note [http://www.w3.org/TR/ws-arch].

[6] Berners-Lee, T., Hendler, J. and Lassila, O. (2001) 'The Semantic Web. *Scientific American* **284**(5), 34–43.

[7] Martin, D., Burstein, M., Hobbs, J., Lassila, O., McDermott, D., McIlraith, S., Narayanan, S., Paolucci, M., Parsia, B., Payne, T., Sirin, E., Srinivasan, N. and Sycara, K. (2004) *Owl-s: Semantic Markup for Web Services*. Member Submission, W3C. World Wide Web Consortium.

[8] Open Mobile Alliance (OMA) (2004) *OMA Web Services Enabler (OWSER): Overview*.

[9] ETSI-3GPP (2005) *Universal Mobile Telecommunications System (UMTS); Open Service Access (OSA); Parlay X Web Services; Part 4: Short Messaging (3GPP TS 29.199-04 version 6.3.0 Release 6)*. Technical Specification.

[10] Colgrave J. and Januszewski, K. (undated) *Using WSDL in a UDDI Registry, version 2.0.2* Technical Note. OASIS.

[11] Dean, M., Schreiber, G., Bechhofer, S., van Harmelen, F., Hendler, J., Horrocks, I., McGuinness, D.L., Patel-Schneider, P.F. and Stein, L.A. (2004) *OWL Web Ontology Language Reference, 2004*. W3C Recommendation. World Wide Web Consortium [http://www.w3.org/TR/owl-ref].

[12] Baader, F., Calvanese, D., McGuinness, D., Nardi, D. and Patel-Schneider, P. (2003) *The Description Logic Handbook – Theory, Implementation and Applications*. Cambridge University Press.

[13] Paolucci, M., Ankolekar, A., Srinivasan, M. and Sycara, K. (2003) 'The DAML-S Virtual Machine' *Proceedings of the Second International Semantic Web Conference*, Sanibel Island, FL.

[14] Masuoka, R., Labrou, Y., Parsia, B. and Sirin, E. (2003) 'Ontology-enabled Pervasive Computing Applications' *IEEE Intelligent Systems* 18(5), 68–72.

[15] Lawrence, S. (2002) *Basic Device Definition version 1.0*. UPnP Standard.

[16] Paolucci, M., Goix, W., Andreetto, A., Luther, M. and Wagner, M. (2005) 'Representing Services for Mobile Computing using OWL and OWL-S: An Initial Investigation' *Proceedings of Web Service Composition Workshop (wscomps05)*.

[17] Denker, G., Kagal, L., Finin, T., Paolucci, M., Srinivasan, N. and Sycara, K. (2003) 'Security for DAML Web services: Annotation and Matchmaking' *Proceedings of the Second International Semantic Web Conference (ISWC 2003)*.

[18] Paolucci, M., Kawamura, T., Payne, T.R. and Sycara, K. (2002) 'Semantic Matching of Web Services Capabilities' *Proceedings of the First International Semantic Web Conference*.

[19] Sycara, K., Paolucci, M., Anolekar, A. and Srinivasan, N. (2003) 'Automated Discovery, Interaction and Composition of Semantic Web Services' *Web Semantics* 1(1).

[20] Li L. and Horrocks, I. (2003) 'E-commerce: A Software Framework for Matchmaking based on Semantic Web Technology' *Proceedings of the Twelfth International Conference on World Wide Web*, Budapest, Hungary.

[21] Mandell, D. and McIllraith, S. (2003) 'A Bottom-up Approach to Automating Web Service Discovery, Customization, and Semantic Translation' *Proceedings of the 12th International Conference on the World Wide Web (WWW 2003)*. ACM Press.

[22] Colucci, S., Noia, T.D., Sciascio, E.D., Donini, F. and Mongiello, M. (2004) 'Concept Abduction and Contraction for Semantic-based Discovery of Matches and Negotiation Spaces in an e-Marketplace' *Proceedings of the 6th International Conference on Electronic Commerce (ICEC 2004)*. ACM Press.

[23] Constantinescu, I. and Faltings, B. (2003) 'Efficient Matchmaking and Directory Services' *Proceedings of IEEE/WIC International Conference on Web Intelligence*.

[24] Klein, M. and Koenig-Ries, B. (2004) 'Coupled Signature and Specification Matching for Automatic Service Binding' *Procedings of European Conference on Web Services*, LNAI, pp. 183–197. Springer.

[25] Banaei-Kashani, F., Chen, C.-C. and Shahabi, C. (2004) 'Wspds: Web services Peer-to-peer Discovery Service' *Proceedings of International Symposium on Web Services and Applications (ISWS)*.

[26] Mallya, A.U., Desai, N., Chopra, A.K. and Singh, M.P. (2005) 'Owl-p: OWL for Protocols and Processes' *Proceedings of the Fourth International Conference on Autonomous Agents and MultiAgent Systems*.

[27] Peer, J. (2005) 'Semantic Service Markup with Sesma' *Proceedings of the Web Service Semantics Workshop (WSS'05) at WWW'05*.

[28] Confalonieri, R., Domingue, J. and Motta, E. (2004) 'Orchestration of Semantic Web services in IRS-III' *Proceedings of the First AKT Workshop on Semantic Web Services (AKT-SWS'04)*, Milton Keynes, UK, December.

[29] Hakimpour, F., Confalonieri, R., Sell, D. and Domingue, J. (2005) 'Orchestration of WSMO-based Semantic Web Services in IRS-III' *Proceedings of the 2nd European Semantic Web Conference (ESWC'05)*, Heraklion, Greece.

[30] Bouquet, P., Giunchiglia, F., van Harmelen, F., Serafini, L. and Stuckenschmidt, H. (2003) 'C-owl: Contextualizing Ontologies' *Second International Semantic Web Conference*, Sanibel Island, FL.

[31] Fensel, D. and Bussler, C. (2002) 'The Web Service Modeling Framework (WSMF)' *Electronic Commerce: Research and Applications* 1(2), 113–137.

[32] Akkiraju, R., Farrell, J., Miller, J., Nagarajan, M., Schmidt, M.-T., Sheth A., Verma, K. (2005) Web Service Semantics—WSDL-S—Version 1.0 Technical Note, April 2005, World Wide Consortium [http://www.w3.org/Submission/WSDL-S/]

[33] Kopecký, Jacek, Moran, Matthew, Vitvar, Tomas, Roman, Dumitru and Mocan, Adrian (undated) *WSMO Grounding* [http://www.wsmo.org/TR/d24/d24.2/v0.1].

[34] Paolucci, Massimo, Wagner, Matthias and Martin, David (2007) 'Grounding OWL-S in SAWSDL' *Proceedings of the Fifth International Conference on Service-Oriented Computing*, Vienna, Austria, September, pp. 416–421.

[35] Martin, David, Paolucci, Massimo and Wagner, Matthias (2007) 'Bringing Semantic Annotation to Web services: OWL-S from the SAWSDL Perspective' *Proceedings of the 6th International Semantic Web Conference*, Busan, Korea, November.

[36] Chen, H., Perich, F., Finin, T. and Joshi, A. (2004) 'SOUPA: Standard Ontology for Ubiquitous and Pervasive Applications' *International Conference on Mobile and Ubiquitous Systems: Networking and Services*, pp. 258–267, Boston, MA, August.

[37] Chen, H., Perich, F., Chakraborty, D., Finin, T. and Joshi, A. (2004) 'Intelligent Agents meet Semantic Web in a Smart Meeting Room' *Proceedings of the Third International Joint Conference on Autonomous Agents & Multi Agent Systems (AAMAS'04)*, New York City, NY, July.

[38] Mrohs, B., Luther, M., Vaidya, R., Wagner, M., Steglich, S., Kellerer, W. and Arbanowski, S. (2005) 'Owl-sf – A Distributed Semantic Service Framework' *Proceedings of Workshop on Context Awareness for Proactive Systems (CAPS 2005)*, pp. 67–77.

[39] Gandon, F. and Sadeh, N. (2004) 'Semantic Web Technologies to Reconcile Privacy and Context Awareness' *Web Semantics Journal* **1**(3).

[40] Noppens, O., Liebig, Th., Schmidt, P., Luther, M. and Wagner, M. 'MobiXPL – A SVG-based Mobile User Interface for Semantic Service Discovery' *Proceedings of the 5th International Conference on Scalable Vector Graphics (SVGOPEN'07)*, Tokyo, Japan, September.

[41] Belecheanu, R., Jawaheer, G., Hoskins, A., McCann, J.A. and Payne, T. (2004) 'Semantic Web Meets Autonomic Ubicomp' *Proceedings of the 3rd International* Semantic Web Conference.

[42] Vaculin, Roman and Sycara, Katia (2007) 'Specifying and Monitoring Composite Events for Semantic Web Services' *The 5th IEEE European Conference on Web Services*. IEEE Computer Society.

[43] Andrews, T., Curbera, F., Dholakia, H., Goland, Y., Klein, J., Leymann, F., Liu, K., Roller, D., Smith, D., Thatte, S., Trickovic, I. and Weerawarana, S. (2003) *Specification: Business Process Execution Language for Web Services, version 1.1* [http://www.ibm.com/developerworks/library/ws-bpel].

[44] Burdett, Kavantzas, N. (2004) WS Choreography Model Overview. W3C Working Draft. World Wide Web Consortium. [http://www.w3.org/TR/ws-chor-model/]

[45] Sirin, Evren, Parsia, Bijan, Wu, Dan, Hendler, James and Nau, Dana (2004) 'HTN Planning for Web Service Composition Using SHOP2' *Journal of Web Semantics* **1**(4), 377–396.

[46] Ghallab, M., Nau, D. and Traverso P. (2004) *Automated Planning*. Elsevier.

Chapter 9: Dynamic Adaptation – Changing Services at Run Time

References

[1] Filman, Robert E., Friedman, Daniel P. (2005) 'Aspect-Oriented Programming Is Quantification and Obliviousness' in Filman, Robert E., Elrad, Tzilla, Clarke, Siobhan, Akşit, Mehmet (eds.), *Aspect-Oriented Software Development*, pp. 21–35, Addison-Wesley.

[2] Filman, Robert E. (2001) 'What is Aspect-Oriented Programming, Revisited' ECOOP 2001 Workshop on Advanced Separation of Concerns [http://trese.cs.utwente.nl/Workshops/ecoop01asoc/papers/Filman.pdf].

[3] Kiczales, Gregor, Lamping, John, Mendhekar, Anurag, Maeda, Chris, Lopes, Cristina, Loingtier, Jean-Marc and Irwin, John (1997) 'Aspect-Oriented Programming' in Akşit, Mehmet and Matsuoka, Satoshi (eds), *Proceedings of the European Conference on Object-Oriented Programming*, LNCS 1241, pp. 220–242. Springer.

[4] Videira Lopes, Cristina (2002) *Aspect-Oriented Programming: An Historical Perspective (What's in a Name?)*. Report UCI-ISR-02-5, University of California, Irvine, CA.

[5] Parnas, David L. (1972) 'On the Criteria to be used in Decomposing Systems into Modules' *Communications of the ACM* **15**(12), 1053–1058.

[6] Pree, Wolfgang (1995) *Design Patterns for Object-Oriented Software Development*. Addison-Wesley.

[7] Ernst, Erik (2003) 'Separation of Concerns' *AOSD 2003 Workshop on Software-Engineering Properties of Languages for Aspect Technologies (SPLAT)*}, Boston, MA, March.

[8] Gosling, James, Joy, Bill, Steele, Guy and Bracha, Gilad (2000) *The Java Language Specification* (2nd edn). Addison-Wesley.

 [9] Kniesel, Günter, Costanza, Pascal and Austermann, Michael (2001) 'JMangler – A Framework for Load-Time Transformation of {Java} Class Files' *First IEEE International Workshop on Source Code Analysis and Manipulation (SCAM 2001)*, Florence, Italy, November. IEEE Computer Society Press [http://www.informatik. uni-bonn.de/\~costanza/SCAM_jmangler.pdf]

[10] Hirschfeld, Robert (2003) 'AspectS – Aspect-Oriented Programming with Squeak' In Akšit, Mehmet, Mezini, Mira and Unland, Rainer (eds), *Objects, Components, Architectures, Services, and Applications for a Networked World*, LNCS 2591, pp. 216–232. Springer.

[11] Brant, John, Foote, Brian, Johnson, Ralph E. and Roberts, Don (1998) 'Wrappers to the Rescue' *Proceedings of the European Conference on Object-Oriented Programming*, LNCS 1445, pp. 396–417. Springer.

[12] Maes, Pattie (1987) *Computational Reflection*. Artificial Intelligence Laboratory, University of Brussels (VUB).

[13] Kiczales, Gregor, des Rivieres, Jim and Bobrow, Daniel G. (1991) *The Art of the Metaobject Protocol*. Addison-Wesley.

[14] Mendhekar, Anurag, Kiczales, Gregor and Lamping, John (1997) *RG: A Case-Study for Aspect-Oriented Programming* Report SPL97-009 P9710044. Xerox PARC.

[15] Lopes, Cristina Videira (1997) *D: A Language Framework for Distributed Programming* Dissertation, College of Computer Science, Northeastern University [http://www.parc.xerox.com/csl/groups/sda/pubs/papers/Lopes-Thesis/dissertation.pdf].

[16] Kay, Alan (2002) *Is Software Engineering an Oxymoron?* Viewpoints Research Institute, Glendale, CA.

[17] Czarnecki, Krzysztof (1998) *Generative Programming: Principles and Techniques of Software Engineering Based on Automated Configuration and Fragment-Based Component Models* PhD Thesis, Technical University of Ilmenau.

[18] Goldberg, Adele and Robson, David (1983) *Smalltalk-80: The Language and its Implementation*. Addison-Wesley.

[19] Hirschfeld, Robert and Wagner, Matthias (2002) 'PerspectiveS – AspectS with Context' *OOPSLA 2002 Workshop on Engineering Context-Aware Object-Oriented Systems and Environments (ECOOSE)*, Seattle, WA, November.

[20] Hirschfeld, Robert, Costanza, Pascal and Yierstrast, Oscar (2008) 'Context-oriented Programming' *Journal of Object Technology* April 2008, www.jot.fm.

Further Reading

Aßmann, Uwe (2003) *Invasive Software Composition*. Springer.

Bracha, Gilad and Cook, William (1990) 'Mixin-based *Inheritance*' *Proceedings of the European Conference on Object-Oriented Programming on Object-Oriented Programming Systems, Languages, and Applications*, Ottawa, ON, Canada, pp. 303–311. ACM Press [http://doi.acm.org/10.1145/97945.97982]

Canning, Peter, Cook, William, Hill, Walter, Olthoff, Walter and Mitchell, John C. (1989) 'F-bounded Polymorphism for Object-Oriented Programming' *Proceedings of the International Conference on Functional Programming Languages and Computer Architecture*, London, September, pp. 273–280. ACM Press [http://doi.acm. org/10.1145/99370.99392]

Clarke, Siobhàn and Walker, Robert (2002) 'Towards a Standard Design Language for AOSD' In Kiczales, Gregor (ed.), *Proceedings of the International Conference on Aspect-Oriented Software Development (AOSD 2002)*, Enschede, The Netherlands, April, pp. 113–119. ACM Press.

Dijkstra, Edsger W. (1976) *A Discipline of Programming*. Prentice-Hall.

Ernst Erik and Lorenz, David H. (2003) 'Aspects and Polymorphism in AspectJ' In Akšit, Mehmet *Proceedings of the International Conference on Aspect-Oriented Software Development (AOSD 2003)*, Boston, MA, March, pp. 150–157. ACM Press.

Gamma, Erich, Helm, Richard, Johnson, Ralph and Vlissides, John (1995) *Design Patterns: Elements of Reusable Object-Oriented Software*. Addison-Wesley.

Hanenberg, Stefan, Hirschfeld, Robert and Unland, Rainer (2003) 'Aspect Weaving: Using the Base Language's Introspective Facilities to Determine Join Points' *Workshop on Advancing the State-of-the-Art in Runtime Inspection (ECOOP 2003)*, Darmstadt, Germany, July [http://www.st.informatik.tu-darmstadt.de/pages/workshops/ ASARTI03/HanenbergASARTI03.pdf].

Hanenberg, Stefan and Unland, Rainer (2001) 'Using and Reusing Aspects in AspectJ' *Workshop on Advanced Separation of Concerns in Object-Oriented Systems (OOPSLA 2001)*, Tampa, FL, October, pp 80–89 [http://www.cs.ubc.ca/\~kdvolder/Workshops/OOPSLA2001/submissions/11-hanenberg.pdf]

Hanenberg, Stefan and Unland, Rainer (2002) 'Roles and Aspects: Similarities, Differences, and Synergetic Potential' In Bellahsene, Zohra, Patel, Dilip and Rolland, Colette (eds), *Object-Oriented Information Systems* LNCS 2425, pp. 507–521. Springer.

Ingalls, Dan, Kaehler, Ted, Maloney, John, Wallace, Scott and Kay, Alan (1997) 'Back to the Future: The Story of Squeak, a Practical Smalltalk Written in Itself' *Proceedings of the Conference on Object-Oriented Programming, Systems, Languages, and Applications*, Atlanta, GA, October, pp. 318–326. ACM Press [http://doi.acm.org/10.1145/263698.263754].

Holland, Ian M. (1992) 'Specifying Reusable Components using Co*ntracts' Proceedings of the European Conference on Object-Oriented Programming*, Utrecht, The Netherlands, LNCS 615, pp. 287–308. Springer.

Jézéquel, Jean-Marc, Plouzeau, Noël, Weis, Torben and Geihs, Kurt (2002) 'From Contracts to Aspects in UML Designs' In Aldawud, Omar, Booch, Grady, Clarke, Siobhàn, Elrad, Tzilla, Harrison, Bill, Kandi, Mohamed and Strohmeier, Alfred (eds), *Workshop on Aspect-Oriented Modeling with UML (AOSD 2002)*, Enschede, The Netherlands, March [http://lglwww.epfl.ch/workshops/aosd-uml/Allsubs/jean.pdf].

Johnson, Ralph and Foote, Brian (1988) 'Designing Reusable Classes' *Journal of Object-Oriented Programming* **1**(2), 25–35.

Keene, Sonya E. (1989) *Object-Oriented Programming in Common Lisp: A Programmer's Guide to CLOS*. Addison-Wesley.

Kiczales, Gregor, Hilsdale, Erik, Hugunin, Jim, Kersten, Mik, Palm, Jeffrey and Griswold, William G. (2001) 'An overview of AspectJ' In Lindskov Knudsen, J. (ed.), *Proceedings of the European Conference on Object-Oriented Programming*, LNCS 2072, pp. 327–353. Springer.

Kristensen, Bent B. and Østerbye, Kasper (1996) 'Roles: Conceptual Abstraction Theory and Practical Language Issues' *Theory and Practice of Object Systems* **2**(3), 143–160.

Lopes, Cristina Videira, Dourish, Paul, Lorenz, David H. and Lieberherr, Karl (2003) 'Beyond AOP: Toward Naturalistic Programming' *Companion of the Conference on Object-Oriented Programming, Systems, Languages, and Applications*, Anaheim, CA, pp. 198–207. ACM Press [http://doi.acm.org/10.1145/949344.949400].

Mezini, Mira and Ostermann, Klaus (2002) 'Integrating Independent Components with On-demand Remodularization' *Proceedings of Conference on Object-Oriented Programming, Systems, Languages, and Applications*, Seattle, WA, pp. 52–67. ACM Press [http://doi.acm.org/10.1145/582419.582426].

Orleans, Doug and Lieberherr, Karl (2001) 'DJ: Dynamic Adaptive Programming in Java' In Yonezawa, Akinori and Matsuoka, Satoshi (eds), *International Conference on Metalevel Architectures and Separation of Crosscutting Concerns (Reflection 2001)*, Kyoto, Japan, September, LNCS 2192, pp. 73–80. Springer.

Pulvermüller, Elke, Speck, Andreas and Rashid, Awais (2000) 'Implementing Collaboration-based Designs using Aspect-Oriented Programming' *Proceedings of TOOLS-USA*, Santa Barbara, CA, July, pp. 95–104.

Smaragdakis, Yannis and Batory, Don (1998) 'Implementing Reusable Object-Oriented Components' *Proceedings of the International Conference on Software Reuse*, Victoria, BC, Canada, June, pp. 36–45. IEEE Computer Society Press [http://citeseer.nj.nec.com/article/smaragdakis98implementing.html]

VanHilst, Michael and Notkin, David (1996) 'Using Role Components to Implement Collaboration-Based Designs' *Proceedings of Conference on Object-Oriented Programming, Systems, Languages, and Applications*, San Jose, CA, October, pp. 359–369. ACM Press [http://doi.acm.org/10.1145/236337.236375, [http://citeseer.nj.nec.com/vanhilst96using.html].

Vlissides, John M. (1996) 'Protection, Part I: The Hollywood Principle' *C++ Report*, February [http://www.squeak.org].

Chapter 10: Context-aware Mobility Management

References

[1] Wei, Q., Farkas, K., Mendes, P., Prehofer, C., Plattner, B. and Nafisi, N. (2003) 'Context-aware Handover Based on Active Network Technology' *IWAN 2003*. Springer.

[2] Prehofer, Christian, Kellerer, Wolfgang, Hirschfeld, Robert, Berndt, Hendrik and Kawamura, Katsuya (2002) 'An Architecture Supporting Adaptation and Evolution in Fourth Generation Mobile Communication Systems' *Journal of Communications and Networks* **4**(4).

[3] Prehofer, C. and Wei, Q. (2002) 'Active Networks for 4G Mobile Communication: Motivation, Architecture and Application Scenarios' *IWAN 2002*. Springer.

[4] Kempf, J. (2001) 'Dormant Mode Host Alerting ("IP paging") problem statement' RFC3132, June.

[5] Castelluccia, C. (2001) 'Extending Mobile IP with Adaptive Individual Paging: A Performance Analysis' *ACM Mobile Computing and Communications Review (MC2R)*, April.

[6] Hsiao-Kuang Wu, Ming-Hui Jin, Jorng-Tzong Horng and Cheng-Yi Ke (2001) 'Personal Paging Area Based On Mobile's Moving Behaviours' *IEEE INFOCOM*, May.

[7] Lei, Z., Sarazdar, C.U. and Mandayam, N.B. (1999) 'Mobility Parameter Estimation for the Optimization of Personal Paging Areas in PCS/Cellular Mobile Networks' *Proceedings of the 2nd IEEE Signal Processing Workshop on Signal Processing Advances in Wireless Communications (SPAWC'99)*, 9–12 May.

[8] Stemm, M. and Katz, R. (1998) 'Vertical Handoffs in Wireless Overlay Networks' *ACM Journal on Mobile Networks and Applications* **3**(4).

[9] Pahlavan, K., Krishnamurthy, P., Hatami, A., Ylianttila, M., Makela, J.P., Pichna, R., Vallstron J. (2000) 'Handoff in Hybrid Mobile Data Networks' *IEEE Communication Magazine*, April.

[10] Chan, P.M.L., Sheriff, R.E., Conforto, P., Tocci, C. (2001) 'Mobility Management Incorporating Fuzzy Logic for a Heterogeneous IP Environment' *IEEE Communication Magazine*, December.

[11] Kounavis, Michael E., Campbell, Andrew T., Ito, G. and Bianchi, G. (2001) 'Design, Implementation and Evaluation of Programmable Handoff in Mobile Networks' *Mobile Networks and Applications* **6**, 443–461.

[12] Prehofer, C., Nafisi, N. and Wei, Q. (2003) 'A Framework for Context-aware Handover Decisions' *IEEE International Symposium on Personal, Indoor and Mobile Radio Communications*, Beijing, China, September.

[13] *Composite Capability/Preference Profiles (CC/PP)* (2004) W3C Working Draft. World Wide Web Consortium [http://www.w3.org/Mobile/CCPP].

[14] Psounis, K. (1999) 'Active Networks: Applications, Security, Safety, and Architectures' *IEEE Communications Surveys*, First Quarter.

[15] Mendes, Paulo, Prehofer, Christian and Wei, Qing (2003) 'Context Management with Programmable Mobile Networks' *IEEE Computer Communication Workshop*.

[16] Prehofer, C. and Wei, Q. (2002) 'Active Networks for 4G Mobile Communication: Motivation, Architecture and Application Scenarios' *International Working Conference on Active Networks*, Zurich, Switzerland.

[17] Keller, R., Ruf, L., Guindehi, A. and Plattner, B. (2002) 'PromethOS: A Dynamically Extensible Router Architecture Supporting Explicit Routing' *IWAN 2002*, December, Springer.

[18] Bossardt, Matthias, Hoog Antink, Roman, Moser, Andreas and Plattner, Bernhard (2003) 'Chameleon: Realizing Automatic Service Composition for Extensible Active Routers' *Proceedings of Fifth Annual International Working Conference on Active Networks (IWAN 2003)*, Kyoto, Japan, December, LNCS. Springer.

[19] Wang, Helen J., Katz, Randy H. and Giese, Jochen (1999) 'Policy-enabled Handoffs across Heterogeneous Wireless Networks' *WMCSA 99*, February. IEEE.

[20] Dey, A. (2000) *Providing Architectural Support for Building Context-Aware Applications* PhD thesis, College of Computing, Georgia Institute of Technology.

[21] Chen, G. and Kotz, D. (2002) 'Context Aggregation and Distribution in Ubiquitous Computing Systems' *Proceedings of IEEE Workshop on Mobile Computing Systems and Applications*, Callicoon, NY, February.

Further Reading

Anetd: Active Networks Daemon (undated) ACTIVE project. ISI & SRI. [http://www.sdl.sri.com/projects/activate/anted].

MGEN – Multi-Generator Toolset (undated) [http://mgen.pf.itd.nrl.navy.mil].

Chapter 11: Contextual Intelligence

References

[1] Mrohs, B., Luther, M., Vaidya, R., Wagner, M., Steglich, S., Kellerer, W. and Arbanowski, S. (2005) 'OWL-SF – a Distributed Semantic Service Framework' *Proceedings of the Workshop on Context Awareness for Proactive Systems (CAPS'05)*, Helsinki, Finland, June, pp. 67–77.

[2] McGuinness D. and van Harmelen, F. (2004) *OWL Web Ontology Language Overview* W3C Recommendation. Wide Web Consortium.

[3] Sameshima, S., Suzuki, J. (2004) *Platform Independent Model and Platform Specific Model for SDOs* Final recommended specification. OMG.

[4] Fielding. T.R. (2000) *Architectural Styles and the Design of Network-based Software Architectures.* PhD thesis, University of California, Irvine.

[5] Mrohs, B., Luther, M. and Vaidya, R. (2005) 'Context-aware Presence Management' *Proceedings of the Workshop on Context Awareness for Proactive Systems (CAPS'05)*, June.

[6] Dey, A.K. (2000) *Providing Architectural Support for Building Context-aware Applications.* PhD thesis, Georgia Institute of Technology.

[7] Chen, H., Finin, T. and Joshi, A. (2004) 'A Context Broker for Building Smart Meeting Rooms. *Proceedings of the Autonomous Systems Symposium*, AAAI Spring Symposium. CA, March.

[8] Khushraj, D. and Lassila, O. (2004) 'CALI: Context Awareness via Logical Inference' *Proceedings of the Workshop on Semantic Web Technology for Mobile and Ubiquitous Applications*, November.

[9] ESSI WSML Working Group. Web Services modeling Language WSML. http://www.wsmo.org/wsml

[10] MacGregor, R. (1991) 'Using a Description Classifier to Enhance Deductive Inference' *Proceedings of the 7th IEEE Conference on AI Applications*, pp. 141–147.

[11] Grosof, B.N., Horrocks, I., Volz, R. and Decker, Stefan (2003) 'Combining Logic Programs with Description Logic' *Proceedings of the 12th International World Wide Web Conference.* ACM.

[12] Fensel, D. and Bussler, C. (2005) 'The Web Service Modeling Framework WSMF' *International Journal of Electronic Commerce* **9**(2).

[13] Paolucci, M., Goix, W., Andreetto, A., Luther M. and Wagner M. (2005) 'Representing Services for Mobile Computing using OWL and OWL-S: An Initial Investigation' *Proceedings of the Workshop on Web service Composition in conjunction with the International Conference on Web Intelligence and Intelligent Agent Technology*, Compiège, France, September 2005.

[14] Beydoun, G. and Hoffmann, A. (2000) 'Monitoring Knowledge Acquisition, Instead of Evaluating Knowledge Bases' *Proceedings of the European Knowledge Acquisition Conference (EKAW2000)*, Juan-les-Pins, France, LNCS 1937. Springer.

[15] W3C. Web Ontology Language (OWL) http://www.w3.org/2004/OWL

Further Reading

3GPP (2002) *The Third Generation Partnership Project* [http://www.3gpp.org].

Balke, W.-T. and Wagner. M. (2003) 'Cooperative Discovery for User-centered Web Service Provisioning' Proceedings of the 1st International Conference on Web Services (ICWS'03), Las Vegas.

Balke, W.-T. and Wagner, M. (2003) 'Towards Personalized Selection of Web Services' *Proceedings of the International World Wide Web Conference (WWW)*, Budapest, Hungary.

Balke, W.-T. and Wagner, M. (2004) 'Through Different Eyes – Assessing Multiple Conceptual Views for Querying Web Services' Proceedings of the 13th International. World Wide Web Conerence. (WWW2004) Alternate Track on Web Services, New York.

Banerji, A., Bartolini, C., Beringer, D., Chopella, V., Govindarajan, K., Karp, A., Kuno, H., Lemon, M., Pogossiants, G., Sharma, S. and Williams, S. (2002) *Web Services Conversation Language (WSCL) 1.0* [http://www.w3.org/TR/wscl10].

Berners-Lee, T., Hendler, J. and Lassila. O. (2001) 'The Semantic Web' *Scientific American*, **284**(5), 34–43.

Casati, F. and Shan. M. (2001) 'Dynamic and Adaptive Composition of E-Services' *Journal of Information Systems* **6**, 143–163.

Chu, H., Yang, K., Chiang, M. Minock, Chow, G. and Larson, C. (1996) 'CoBase: A Scalable and Extensible Cooperative Information System' *Journal of Intelligent Information Systems (JIIS)* **6**(3), 223–259.

IETF and World Wide Web Consortium (2002) *XML Signature* [http://www.w3.org/Signature].

Kießling, W. and Köstler, G. (2002) 'Preference SQL – Design, Implementation, Experiences' *Proceedings of the International Conference on Very Large Data Bases (VLDB'02)*, Hong Kong, China.

Leymann, F. (2001) *Web Services Flow Language (WSFL 1.0)* [http://www-4.ibm.com/software/solutions/webservices/pdf/WSFL.pdf].

Luther, M., Mrohs, B., Wagner, M., Steglich, S. and Kellerer, W. (2005) 'Situational Reasoning – A Practical OWL Use Case' *Proceedings of the 7th International Symposium on Autonomous Decentralized Systems (ISADS'05)*, Chengdu, China, April.

Minker, J. (1998) 'An Overview of Cooperative Answering in Data-bases' *Proceedings of the Inernational. Conference on Flexible Query Answering Systems (FQAS)*, Roskilde, Denmark, LNCS 1495. Springer.

Motro, A. (1988) 'VAGUE: A User Interface to Relational Databases that Permits Vague Queries' *ACM Transactions on Office Information Systems (TOIS)*, **6**, 187–214.

Narayanan, S. and McIlraith, S. (2002) 'Simulation, Verification and Automated Composition of Web Services' *Proceedings of the International World Wide Web Conference (WWW 2002)*, Honolulu, Hawaii, pp. 77–88.

Paolucci, M., Kawamura, T., Payne, T. and Sycara, K. (2002) 'Semantic Matching of Web Services Capabilities' *Proceedings of the International Semantic Web Conference (ISWC'02)*, Sardinia, Italy.

Paolucci, M., Kawamura, T., Payne, T. and Sycara, K. (2002) 'Importing the Semantic Web in UDDI' *Proceedings of the International Workshop on Web Services, e-Business and the Semantic Web (WES'02)*, Toronto, Canada.

Parlay Group, The (undated) *Parlay/OSA APIs* [http://www.parlay.org].

Pires, P., Benevides, M. and Mattoso, M. (2002) 'Building Reliable Web Services Compositions' *Proceedings of the International Workshop on Web Services: Research, Standardization and Deployment (WS-RSD)*, Erfurt, Germany, pp. 551–562.

Thatte, S. (2001) *XLANG: Web Services for Business Process Design* [http://www.gotdotnet.com/team/xml_wsspecs/xlang-c/default.htm].

Vilain, M. (1990) 'Getting Serious about Parsing Plans: A Grammatical Analysis of Plan Recognition' *Proceedings of the National Conference on Artificial Intelligence (AAAI-90)*, Boston, pp. 190–197.

Wagner, M., Balke, W.-T., Hirschfeld, R. and Kellerer, W. (2002) 'A Roadmap to Advanced Personalization of Mobile Services' *Proceedings of the International Conference DOA/ODBASE/ CoopIS (Industry Program)*, Irvine, CA.

Wagner, M., Kießling, W. and Balke, W.-T. (2002) 'Progressive Content Delivery for Mobile E-Services' *Proceedings of the 3rd International Conferenc. on Advances in Web-Age Information Management (WAIM2002)*, Beijing, China, LNCS 2419, pp. 225–235. Springer.

Chapter 12: From Personal Mobility to Mobile Personality

References

[1] Wagner, M., Luther, M., Hirschfeld, R. , Kellerer, W. and Tarlano, A. (2005) 'From Personal Mobility to Mobile Personality' *Telenor Telektronikk Magazine*, Special Issue on Future Mobile Phone.

[2] Wagner, M., Balke, W.-T., Hirschfeld, R. and Kellerer, W. (2002) 'A Roadmap to Advanced Personalization of Mobile *Services*' *Proceedings of the International Conference DOA/ODBASE/ CoopIS (Industry Program)*, Irvine, CA, October.

[3] Kellerer, W., Wagner, W. and Balke, W.-T. (2003) 'Preference-based Session Management' *Proceedings of the 8th International Workshop on Mobile Multimedia Communications (MOMUC2003)*, Munich, Germany, October.

[4] Lassila, O. and Swick, R.R. (1999) *Resource Description Format: Model and Syntax Specification*. W3C Recommendation. World Wide Web Consortium.

[5] Berners-Lee, T., Hendler, J. and Lassila, O. (2001) 'The Semantic Web' *Scientific American*, **284**(5), 34–43.

[6] McGuinness, D.L. and van Harmelen, F. (2003) *OWL Web Ontology Language Overview*. W3C Working Draft. World Wide Web Consortium [http://www.w3.org/TR/owl-features].

[7] Balke, W.-T. and Wagner, M. (2003) 'Towards Personalized Selection of Web Services' *Proceedings of the International World Wide Web Conference (WWW2003)*, Budapest, Hungary, May.

[8] Balke, W.-T. and Wagner, M. (2003) 'Cooperative Discovery for User-centered Web Service Provisioning' *Proceedings of the 1st International Web Service Conference (ICWS2003)*, Las Vegas, NV.

[9] Naghshineh, M. (2002) 'Context-Aware Computing' *IEEE Wireless Communications Magazine*, Special Issue, October.

[10] Floreen, P., Przybilski, M., Nurmi, P., Koolwaaij, J., Tarlano, A., Wagner, M., Luther, M., Bataille, F., Boussard, M., Mrohs, B. and Lau, S. (2005) 'Towards a Context Management Framework for MobiLife' *Proceedings of the 14th IST Mobile & Wireless Communications Summit*, Dresden, Germany, 19–23 June.

[11] Koolwaaij, Johan, Tarlano, Anthony, Luther, Marko, Nurmi, Petteri, Mrohs, Bernd, Battestini, Agathe and Vaidya, Raju (2006) 'Context Watcher – Sharing Context Information in Everyday Life' *Proceedings of the IASTED International Conference on Web Technologies, Applications, and Services (WTAS2006)*, Calgary, Canada, July.

[12] Kellerer, W. and Berndt, H. (2002) 'Next Generation Service Session Signaling' *Proceedings of TINA 2002*, Petaling Jaya, Malaysia.

[13] Prehofer, C., Kellerer, W., Hirschfeld, R., Berndt, H. and Kawamura, K. (2002) 'An Architecture Supporting Adaptation and Evolution in Fourth Generation Mobile Communication Systems' *Journal of Communications and Networks* **4**(4).

[14] Hirschfeld, R. and Wagner, M. (2002) 'PerspectiveS – AspectS with Context' *Proceedings of the OOPSLA 2002 Workshop on Engineering Context-Aware Object-Oriented Systems and Environments*, Seattle, WA, November.

[15] Hirschfeld, R. (2002) 'AspectS – Aspect-Oriented Programming with Squeak' *Architectures, Services, and Applications for a Networked World*, LNCS 2591. Springer.

[16] Hirschfeld, R., Wagner, M., Kellerer, W. and Prehofer, C. (2003) 'AOSD for System Integration and Personalization' *Proceedings of the AOSD Workshop on the Commercialization of AOSD Technology*, Boston, MA, March.

[17] Noppens, O., Luther, M., Liebig, T., Wagner, M. and Paolucci, M. (2006) 'Ontology-based Preference Handling for Mobile Music Selection' *Proceedings of the 3rd Workshop on Advances in Preference Handling in conjunction with ECAI'06*, Riva del Garda, Italy.

[18] Wagner, M., Liebig, T., Noppens, O., Balzer, S. and Kellerer, W. (2004) 'mobiXPL – Semantic-based Service Discovery on Tiny Mobile Devices' *Proceedings of Workshop on Semantic Web Technology for Mobile and Ubiquitous Applications (in conjunction with ISWC'04)*, Hiroshima, Japan.

[19] Kießling, W. and Köstler, G. (2002) 'Preference SQL – Design, Implementation, Experiences' *Proceedings of the International Confernce on Very Large Databases (VLDB'02)*, Hong Kong.

[20] Balke, W.-T. and Wagner, M. (2004) 'Through Different Eyes – Assessing Multiple Conceptual Views for Querying Web Services' *Proceedings of 13th International World Wide Web Conference (WWW'04)*, New York.

Index

Towards 4G Technologies. Edited by Hendrik Berndt.
© 2008 John Wiley & Sons, Ltd.